Introduction to STRENGTH of Materials

D S PRAKASH RAO

Universities Press

Universities Press (India) Private Limited

Registered Office
3-6-747/1/A & 3-6-754/1, Himayatnagar,
Hyderabad 500 029 (A.P.), India
e-mail: info@universitiespress.com

Distributed by
Orient Blackswan Private Limited

Registered Office
3-6-752, Himayatnagar, Hyderabad 500 029 (A.P.), India

Other Offices
Bangalore / Bhopal / Bhubaneshwar / Chennai
Ernakulam / Guwahati / Hyderabad / Jaipur / Kolkata
Lucknow / Mumbai / New Delhi / Patna

First Published 2002
Reprinted 2009, 2013

ISBN: 978 81 7371 405 4

Typeset by
OSDATA, Hyderabad 500 029

Printed in India at
Graphica Printers
Hyderabad 500 013

Published by
Universities Press (India) Private Limited
3-6-747/1/A & 3-6-754/1, Himayatnagar
Hyderabad 500 029 (A.P), India

Contents

Preface *vii*

Notation *ix*

I Introduction **1**

I.0 Introduction 1

I.1 System of units 1

I.2 Conventions and notation 2

I.2.1 Forces and moments 2

I.2.2 Coordinate system 3

I.2.3 Deformations 4

I.2.4 Supports 4

I.3 External forces on a body 4

I.3.1 Direct forces 6

I.3.2 Transverse forces 6

I.3.3 Equilibrium of a body 6

I.4 Internal forces of a body 8

I.5 Free body diagrams 9

I.6 Summary 9

1 Direct stresses **10**

1.0 Introduction 10

1.1 Stress 10

1.2 Strain 11

1.3 Hooke's law and Young's modulus 12

1.4 Stress–strain behaviour of materials 13

1.4.1 Mild steel 13

1.4.2 High strength steels 14

1.4.3 Other materials 15

1.4.4 Ductility 16

1.4.5 Elastic and plastic ranges 17

1.5 Factor of safety 18

1.6 Elongation of a bar 18

1.7 Principle of superposition 18

1.8 Members with varying cross section 24
1.9 Compound bars 27
1.10 Thermal stresses 34
1.10.1 Thermal stresses in simple bars 34
1.10.2 Thermal stresses in compound bars 35
1.11 Poisson's ratio 39
1.12 Bulk modulus 41
1.13 Summary 43
1.14 Exercises 44

2 Impact loading and strain energy **51**
2.0 Introduction 51
2.1 Impact loading 51
2.2 Work and energy of a system 51
2.2.1 External work 52
2.2.2 Internal work 53
2.2.3 Energy principle 54
2.3 Impact load analysis 55
2.4 Summary 64
2.5 Exercises 65

3 Shear stresses **67**
3.0 Introduction 67
3.1 Types of direct shear 67
3.2 Shear stress 69
3.3 Complementary shear 69
3.4 Shear strain 70
3.5 Rigidity modulus 71
3.6 Strain energy 72
3.7 Summary 72
3.8 Exercises 72

4 Stress analysis **74**
4.0 Introduction 74
4.1 Conventions and notation 74
4.1.1 Direct stresses 74
4.1.2 Shear stresses 74

4.1.3 Inclination of planes 75
4.2 Simple stress condition 75
4.3 Pure shear stress 78
4.4 Bi-axial stress condition 79
4.5 General stress analysis (combined bi-axial and shear stresses) 81
4.6 Relation between Young's modulus and rigidity modulus 85
4.7 Mohr's stress circle 86
4.7.1 Negative shear stresses 88
4.7.2 $\sigma_x < \sigma_y$ 88
4.7.3 $\sigma_x = \sigma_y = 0$ 89
4.7.4 $\sigma_x < 0$ 90
4.8 Summary 91
4.9 Exercises 91

5 Analysis of beams 93
5.0 Introduction 93
5.1 Types of beams 93
5.2 Loading on beams 94
5.3 Forces on a beam section 94
5.3.1 Shear force and B.M. 95
5.3.2 Convention for S.F. and B.M. 95
5.3.3 Bending moment and S.F. diagrams 96
5.4 Relation between load, S.F. and B.M. 97
5.4.1 Relation between load and S.F. 97
5.4.2 Relation between S.F. and B.M. 98
5.5 Summary 110
5.6 Exercises 111

6 Flexural analysis of beams 112
6.0 Introduction 112
6.1 Assumptions of bending theory 112
6.2 Euler–Bernoulli's bending equation 113
6.3 Moment of inertia 115
6.3.1 Theorem of parallel axes 115
6.3.2 Theorem of perpendicular axes 116
6.4 Moments of inertia of arbitrary sections 120

6.5 Section modulus 124
6.5 Beams of uniform strength 127
6.7 Summary 127
6.8 Exercises 128

7 Combined stresses **129**
7.0 Introduction 129
7.1 Analysis 129
7.1.1 Eccentricity in one direction 129
7.1.2 Eccentricity in two directions 130
7.2 Kern of a section 132
7.2.1 Rectangular section (middle third rule) 133
7.2.2 Circular section (middle quarter rule) 134
7.3 Summary 135
7.4 Exercises 135

8 Torsional stresses **137**
8.0 Introduction 137
8.1 Torsion analysis 137
8.1.1 Assumptions of torsion analysis 137
8.1.2 Torsion equation 138
8.1.3 Polar M.I. 139
8.1.3.1 Solid circular section 139
8.1.3.2 Hollow circular section 140
8.1.3.3 Thin circular tube 140
8.1.4 Longitudinal stresses in shafts 144
8.2 Strain energy in elastic shafts 145
8.3 Combined torsion and bending 146
8.3.1 Solid circular shafts 147
8.4 Compound shafts 149
8.5 Summary 152
8.6 Exercises 152

Answers to selected problems *155*
Glossary *158*
References *160*
Index *161*

Preface

The theories of strength of materials involve a judicious combination of mathematical analysis and scientific observations with experimental results. The subject deals basically with the behaviour of materials subjected to static loading, and investigates the response of isolated members such as beams, columns, shafts and springs. These elements form components of engineering systems like machines, automobiles, space-crafts, towers and several other structures.

Engineers should be familiar with the load sustaining capacities of the systems they design and use in practical applications. In ancient times, engineers designed mainly war machines and buildings. They relied on the rules of thumb formulated by their predecessors, and supplemented them with their own experiences and intuition. Most of the developments in the theory commenced possibly with Leonardo da Vinci (1452–1549). Galileo Galilei (1564–1642) is generally regarded as the progenitor of mechanics of materials; he conducted several pioneering experiments on the behaviour of beams with solid and hollow sections. He was perhaps the first to note that the strength of members can be increased several times by adopting hollow sections instead of solid sections without increasing the weight. The seventeenth and eighteenth centuries were the most exciting periods in the annals of human civilisation when several theories were propounded and led to the establishment of scientific procedures in analysis and design. Robert Hooke (1635–1703), Mariotte (1620–1684), Jacob Bernoulli (1654–1705), John Bernoulli (1667–1748), Daniel Bernoulli (1700–1782), Euler (1707–1783), Lagrange (1736–1813), Parent (1666–1716), Coulomb (1736–1806) and Navier (1785–1836), among several others, made significant contributions to the development of various theories. It is a far cry from the days of early Asian, Egyptian and European civilisations, when magnificent structures were constructed perhaps without any recourse to theoretical analysis. Even the availability of superfast computers and powerful programme packages these days do not render those theories redundant.

Besides the theories, students should also be familiar with the standard notation, conventions and units of measurement as well as presentation of computations and results. The first book on strength of materials was published in 1826 by Navier, incorporating all his theories and the contributions of the preceding mathematicians and physicists. Since then several excellent books on the subject have been published from various countries from time to time, which formed the basis of the present book. However, the present book, developed from lectures delivered at various institutions, includes the current practices and adopts SI units, standard conventions and notation.

The book is designed to meet the requirements of an elementary course on strength of materials for undergraduate programme in engineering and technology. The introductory chapter helps recapitulate the principles of engineering mechanics and familiarises students with the system of units, notation and conventions adopted. The theory of direct stresses and its applications to prismatic, non-prismatic and compound bars are explained in Chapter 1. Chapter 2 deals with impact loading, and strain energy principles. The concepts of shear stress are discussed in Chapter 3, while stress analysis is explained in Chapter 4. Analysis of determinate beams subjected to bending is presented in Chapter 5 along with the concepts of bending moment and shear force diagrams. Beam sections subjected to bending stresses are analysed in Chapter 6; moment of inertia of various sections and its significance are also explained. Chapter 7 deals with sections subjected to combined stresses (bending moment and axial forces) including the concept of core of a section and its applications. Torsional effects on prismatic, non-prismatic and compound shafts are discussed in Chapter 8. The systematic organisation of the topics enables students to appreciate the basic theories and their applications, and prepares them for dealing with complex systems later. The book forms the basis for other subjects such as machine elements, machine tool design and computer aided design taught in the higher classes of the degree course.

The book presents the topics of an introductory course on the subject; students may refer to the author's books on strength of materials for more topics and advanced analysis.

The efforts of M/s Universities Press, Hyderabad in bringing out this book in an excellent format within a short time to suit the requirements of a condensed course on the subject are appreciated.

It is a pleasure to dedicate this book to my adorable child Dr. Deepa Rao Basu and her Kargil hero Captain (Dr.) Somnath Basu, S M.

Suggestions and comments on the book are welcome from students and professional colleagues.

December 2001 D.S. Prakash Rao

Notation

Roman letters

A	area
a	acceleration, distance
B, b	width
D, d	depth, diameter
E	Young's modulus (elasticity modulus)
e	elongation, eccentricity (subscript indicates the direction)
F	shear force
G	shear modulus
g	acceleration due to gravity
I	moment of inertia (subscripts x and y indicate the respective axes)
J	polar moment of inertia
K	bulk modulus
L, l	length, span
M	bending moment, moment
m	mass
P	load
R	reaction, radius
r	radius
T	temperature, torque
t	thickness
U	strain energy
u	strain energy per unit volume
V	volume
w	load per unit length
x, y	co-ordinates, distances
Z	section modulus

Greek letters

α	coefficient of thermal expansion (subscript indicates the material)
δ	deflection
ε	direct strain (subscript indicates the direction or the material)
γ	shear strain
σ	direct (normal) stress (subscript indicates the direction or the material)
τ	shear stress
ν	Poisson's ratio
ρ	density of the material

Abbreviations

A.F.	axial force	B.M.	bending moment
F.B.	free body	S.F.	shear force

I Introduction

I.0 Introduction

Engineers strive to design systems that are safe, durable and economical. It is essential to know the load carrying capacity of various members of a system, be it a structure or a machine, in order to determine their dimensions for the minimum possible cost. Strength of materials is a subject that deals with the load carrying capacity (strength), rigidity and stability of isolated members such as beams, columns and shafts.

The theory of strength of materials is presented in this book in a systematic way to enable students understand the basic principles and apply them to large structures and systems subsequently. It should be appreciated that even awe inspiring structures, such as heavy machines, cranes, high rise towers and space crafts, rely on these principles for their analysis and design. The system of units, notation and conventions are explained in this chapter. Students are advised to note them before studying the subject itself.

I.1 System of units

Systéme International d'Unités (SI units) are used in this text; the basic units used are gram (g, mass), metre (m, length), second (s, time) and degree Celsius or Kelvin (C or K, temperature). All other units are derived from the basic units. Students should refer standard publications on SI units for precise definitions. Some of the basic concepts of units are included here for ready reference. Force F is derived from Newton's second law of motion, given by

$$F = ma$$

where m = mass of the body and a = acceleration.

If the force on a structure is due to gravity loading, the parameter a is replaced by g (acceleration due to gravity). The basic unit of force is newton (N), which is defined as the force required to impart unit acceleration (1 m s^{-2}) to a mass of 1 kilogram (kg).

Thus, the force due to gravity on a body of 1 kg mass will be

$$F = 1.0\ g\ \text{N} = 9.8066\ \text{N}$$

Since newton is a small quantity, forces are usually expressed in kilo newton (kN) or mega newton (MN). Distributed loads are expressed in kNm^{-1} or kNm^{-2}, depending upon whether the load is distributed over a given length or over a given surface area, respectively.

Stress at a point is expressed as the force per unit area (Nm^{-2}). However, since the unit Nm^{-2} is an extremely small quantity (equivalent to placing a small apple of about 100 g mass on an area of 1.0 × 1.0 m), stress is usually expressed in MNm^{-2}, which is the same as Nmm^{-2}. The unit Nm^{-2} is also referred as pascal (Pa). Similarly, moments are expressed in kNm or MNm (force × distance).

The multiples or fractions of base units are expressed in terms of 10^{+3} or 10^{-3}; other indices such as 10^{-2}, 10^{5} or 10^{7} are seldom used. Some commonly used multiples and fractions of base units are given in Table I.1.

Table I.1 *Multiplication factors for base units*

S.No.	Prefix	Symbol	Factor
1	giga	G	10^{9}
2	mega	M	10^{6}
3	kilo	k	10^{3}
4	milli	m	10^{-3}
5	micro	μ	10^{-6}
6	nano	n	10^{-9}

Note It should be noted that the units of a parameter (force, moment, displacement) should be written in the way indicated in Table I.1. For instance, stress should be written as MNm^{-2} or MPa without hyphens or decimals in between the symbols. Students should note that the symbols are to be followed rigidly, and it is incorrect to denote metres by M instead of m or to denote kNm as kNM.

The units should be mentioned in full and not as abbreviations verbally. For instance, it is incorrect to say mm or kNm while reading or speaking; it should be read and spoken as *milli metres* or *kilo newton metres*.

Numerical values are written in clusters of three digits without commas in between; for instance the correct method of writing is 195 000.0 MPa, and not 1,95,000 MPa or 1 95 000 MPa. Any parameter should be written with adequate number of digits after the decimal; if there are no digits after decimal, it is still necessary to include a zero after decimal as indicated above. Fractional values should be preceded by a zero and decimal. A value should be written as 250.0 mm or 0.25 m, but not as 250. mm or .25 m.

Further, numerical values should be rounded off to the required precision, and not truncated. The value 563.276 m can be written as 563.28 m or as 563.3 m depending upon the precision required. It is incorrect to write the above value as 563.27 m or 563.2 m; the digits should not be deleted without rounding off. The specifications of IS:2–1960 should be followed in any case.

I.2 Conventions and notation

Conventions and notation usually differ from book to book, but the internationally accepted standard conventions and notation are adopted in this text. Notation adopted is listed under a separate head at the beginning of the book, and repeated wherever necessary later. It should be noted that no generality will be lost by adopting any other conventions and notation. However, students are advised to adopt the standard terminology explained here and in later chapters.

I.2.1 Forces and moments

A force is denoted by means of an arrow indicating the direction; its magnitude is also indicated alongside (Fig.I.1(a)). A force acting on a body is often termed as the load on the body; it is also termed as concentrated force (or load) and point load.

A load distributed over a length is known as distributed load (or force). It is indicated by means of a train of arrows over its length; the extent and magnitude of the load are also indicated. If the distributed load is of constant magnitude, it is known as *uniformly distributed load* (u.d.l.). The magnitude of load (or load intensity) may also vary along the length, when it is known as varying load; the variation is termed *uniformly varying load* if the variation is linear. The intensities of load at the ends of the distribution length are also indicated in such a case. This convention is illustrated in Fig.I.1(b).

Moments are indicated by an arc of a circle with the direction shown by an arrow at its end. Couples are indicated by equal and opposite forces; the magnitudes of forces and the distance between them are also indicated. Clockwise moments and couples are considered positive while counter-clockwise moments and couples are negative as indicated in Fig.I.1(c). Tensile forces that tend to elongate a body are termed positive, whereas compressive forces that tend to shorten a body are termed negative.

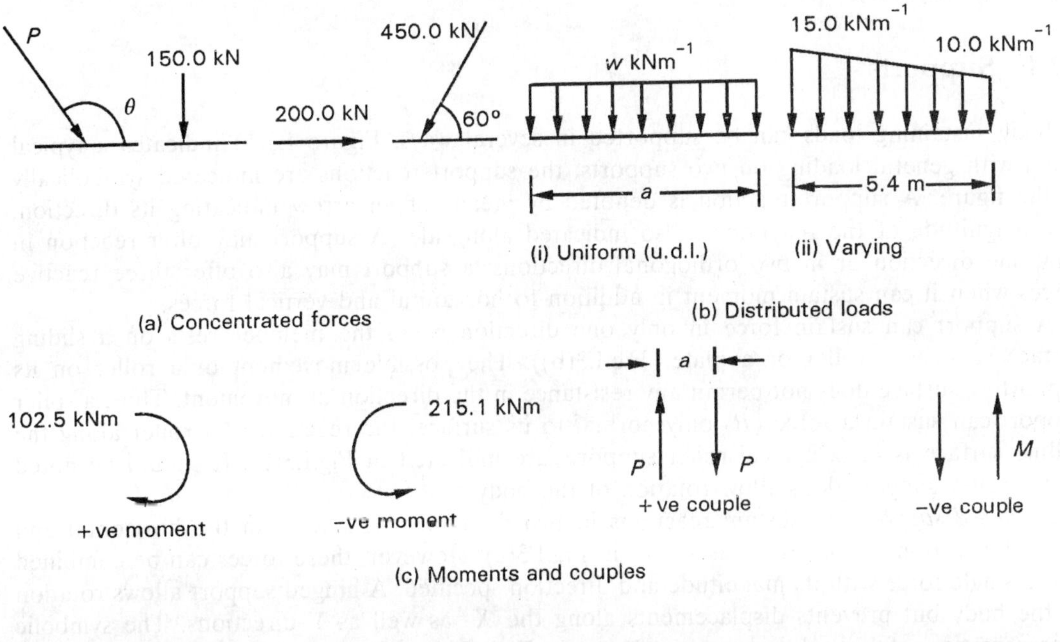

Fig.I.1 Conventions

I.2.2 Coordinate system

The coordinate system adopted is indicated in Fig.I.2(a); the X-axis is taken positive to the right and the Y-axis positive downwards. Horizontal distances measured towards right are positive while those towards the left are negative; vertical distances measured downwards are positive while those measured upwards are negative. Forces acting in the positive directions of the axes are assumed to be positive; that is, horizontal forces acting towards right and vertical forces acting downwards are positive (Fig.I.2(b)). Horizontal forces directed to the left and vertical forces directed upwards are negative (Fig.I.2(c)).

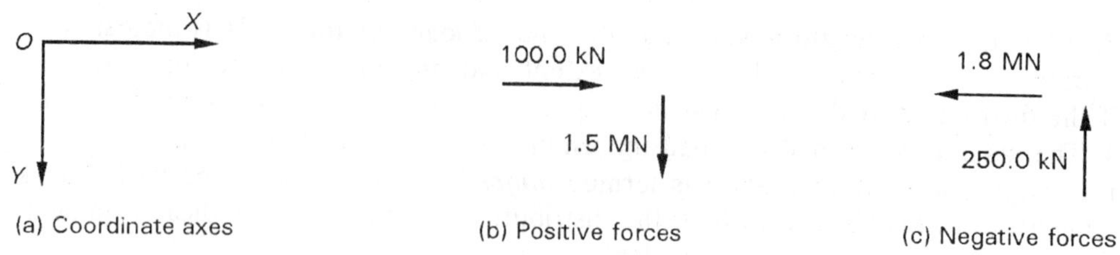

Fig.I.2 Coordinate system

I.2.3 Deformations

Elongation of a body is assumed positive deformation, while shortening is negative deformation. Such a convention is consistent with that for forces (tensile forces are positive and compressive forces are negative).

I.2.4 Supports

A body sustaining loads can be supported in several ways. Figure I.3(a) indicates a typical beam with general loading on two supports; the support reactions are indicated symbolically in the figure. A support reaction is denoted by means of an arrow indicating its direction. The magnitude of the reaction is also indicated alongside. A support may offer reaction in only one direction or in two orthogonal directions; a support may also offer three reactive forces when it can sustain moment in addition to horizontal and vertical forces.

A support can sustain force in only one direction when the member rests on a sliding surface such as a roller or a plate (Fig.I.3(b)). The possible movement of a roller on its supporting surface does not permit any resistance in the direction of movement. Thus, a roller support can sustain a force (R) only normal to its surface; the reaction of a roller along the rolling surface is zero. Typical roller supports are indicated in Fig.I.3(b). It should be noted that a roller support does allow rotation of the body.

A *hinged support* can sustain reactions in two directions; H and R in the horizontal and vertical directions respectively, as shown in Fig.I.3(c). However, these forces can be combined into a single force with its magnitude and direction specified. A hinged support allows rotation of the body but prevents displacements along the X- as well as Y-directions. The symbolic representation of hinged supports is shown in Fig.I.3(c). A body may be built into a rigid support that does not permit any movement including rotation. Such a support can sustain three forces R_1, R_2 and H as indicated in Fig.I.3(d). However, two of these forces (R_1 and R_2) can be combined to form a moment M (couple) and a reaction R. Thus, three reactive forces are sustained at such supports as indicated in the figure. Such a support is known as *fixed* or *built-in* or *encastre support*.

I.3 External forces on a body

A force acting on a body is also referred to as load on the body. According to Newton's first law of motion, force is an entity that changes the state of a body. It should be noted that body is a general term in structural mechanics; it could be a simple bar, a combination of bars, a beam, a column or a frame comprising beams and columns. A body is subjected to both external and internal forces. Any load applied on the boundary (surface) of a body is termed *external force* (or external load), whereas a force developed in the body is termed *internal force*; internal force may be due to external forces or any other causes.

External forces on a body may be due to

- gravity (self-weight of the body itself)
- forces transferred by other bodies through physical contact including reactive forces from supports

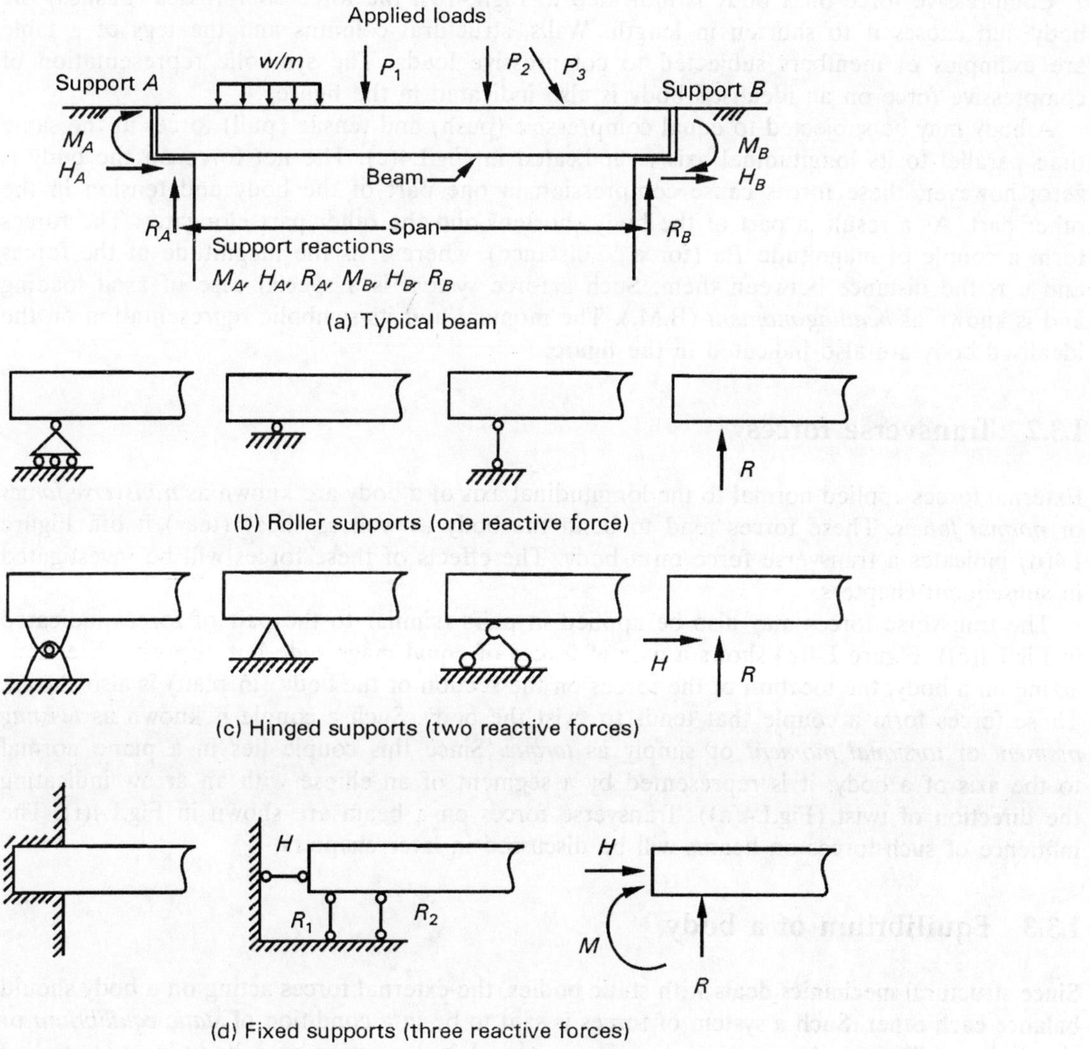

Fig.I.3 Various types of supports

I.3.1 Direct forces

Forces acting along the axis of a body are known as direct forces or axial forces (A.F.). Two types of A.F. should be distinguished:

- axial tensile force (a pull on a body)
- axial compressive force (a push on a body)

Axial forces (or loads) acting on a body are indicated in Fig.I.4. Figure I.4(a) indicates tensile force on a body—the force pulls the body and causes it to elongate. A rope lifting loads or pulling a wagon is subjected to tensile forces, for instance. The symbolic representation of tensile force on a body, idealised as a linear element, is also indicated in the figure.

Compressive force on a body is indicated in Fig.I.4(b); the force compresses (pushes) the body and causes it to shorten in length. Walls, structural columns and the legs of a table are examples of members subjected to compressive loads. The symbolic representation of compressive force on an idealised body is also indicated in the figure.

A body may be subjected to equal compressive (push) and tensile (pull) forces at the same time parallel to its longitudinal axis as indicated in Fig.I.4(c). The net force on the body is zero; however, these forces cause compression in one part of the body and tension in the other part. As a result, a part of the body shortens and the other part elongates. The forces form a couple of magnitude Pa (force $\times$ distance), where P is the magnitude of the forces and a is the distance between them. Such a force system is a special case of axial loading and is known as *bending moment* (B.M.). The moment and its symbolic representation on the idealised body are also indicated in the figure.

I.3.2 Transverse forces

External forces applied normal to the longitudinal axis of a body are known as *transverse forces* or *normal forces*. These forces tend to bend the body as well as shear (tear) it off. Figure I.4(d) indicates a transverse force on a body. The effects of these forces will be investigated in subsequent chapters.

The transverse forces may also be applied in pairs (similar to the pair of forces indicated in Fig.I.4(c)). Figure I.4(e) shows a pair of forces of equal magnitude but opposite directions acting on a body; the location of the forces on the section of the body (in plan) is also shown. These forces form a couple that tends to twist the body. Such a couple is known as *twisting moment* or *torsional moment* or simply as *torque.* Since this couple lies in a plane normal to the axis of a body, it is represented by a segment of an ellipse with an arrow indicating the direction of twist (Fig.I.4(e)). Transverse forces on a beam are shown in Fig.I.4(f). The influence of such forces on beams will be discussed in later chapters.

I.3.3 Equilibrium of a body

Since structural mechanics deals with static bodies, the external forces acting on a body should balance each other. Such a system of forces is said to be in a condition of *static equilibrium* or simply in *equilibrium.* As mentioned earlier, external forces acting on a body include applied forces as well as reactive forces from its supports. Thus, applied forces and reactive forces acting on a body should form a system of forces in equilibrium. These forces together should satisfy the equations of equilibrium indicated below.

$$\sum F_x = \sum F_y = \sum M = 0 \qquad [I.1]$$

The above equations state that the sum of the forces acting on a body in the X-direction, in the Y-direction and the sum of moments of all the forces about any point should be equal to zero.

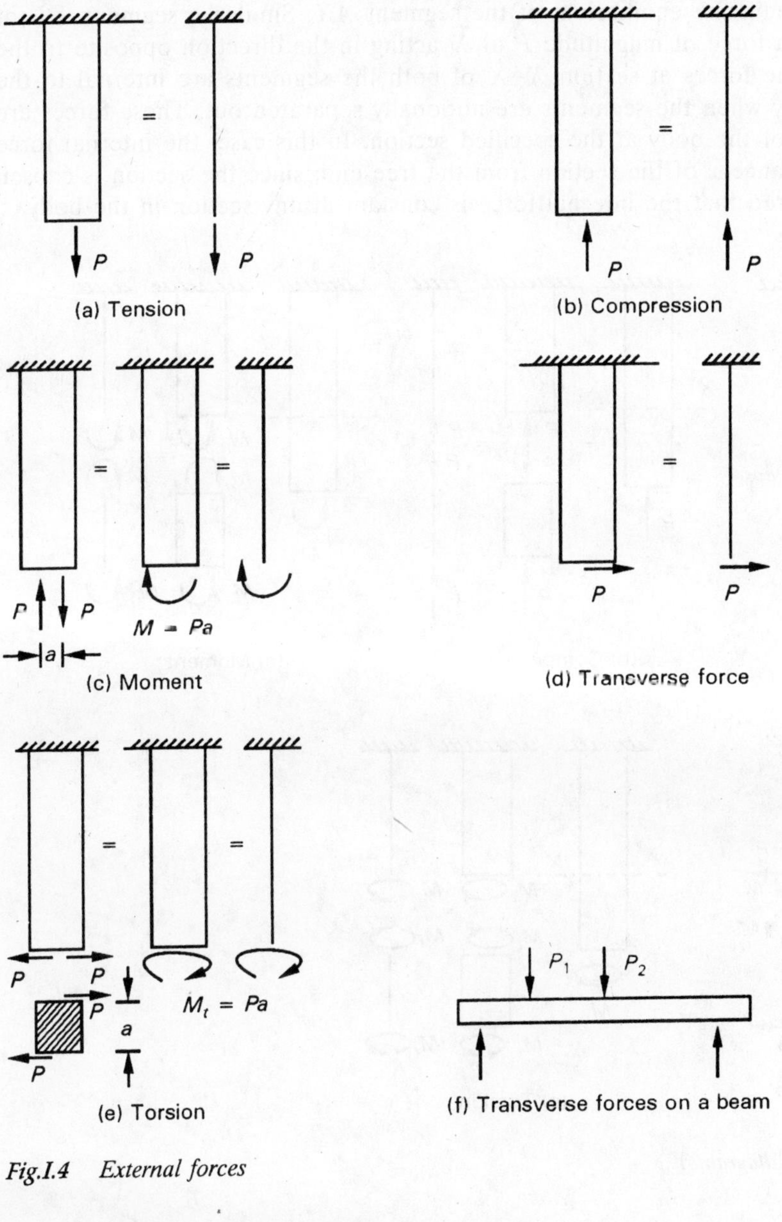

Fig.I.4 External forces

I.4 Internal forces of a body

External forces acting (or applied) on a body are transmitted to its supports through the material of the body. Such a transmission of forces through a body causes the body to deform and develop equal and opposite internal reactions; these are known as *internal forces*.

Figure I.5(a) indicates a body subjected to an external force at its free (unsupported) end. Consider an arbitrary section X–X of the body at a distance x from its free end. If the part of the body below section X–X is separated out, there should be a force of magnitude equal to the applied force P but in the opposite direction at section X–X. The force at section X–X is required to maintain the equilibrium of the segment AX. Similarly, segment XB of the body should develop a force of magnitude P at X acting in the direction opposite to the support reaction at B. The forces at section X–X of both the segments are internal to the body, and are evident only when the segments are notionally separated out. These forces are known as internal forces of the body at the specified section. In this case, the internal force is independent of the distance x of the section from the free end; since the section is chosen arbitrarily, it can be inferred that the internal force is constant at any section in the body.

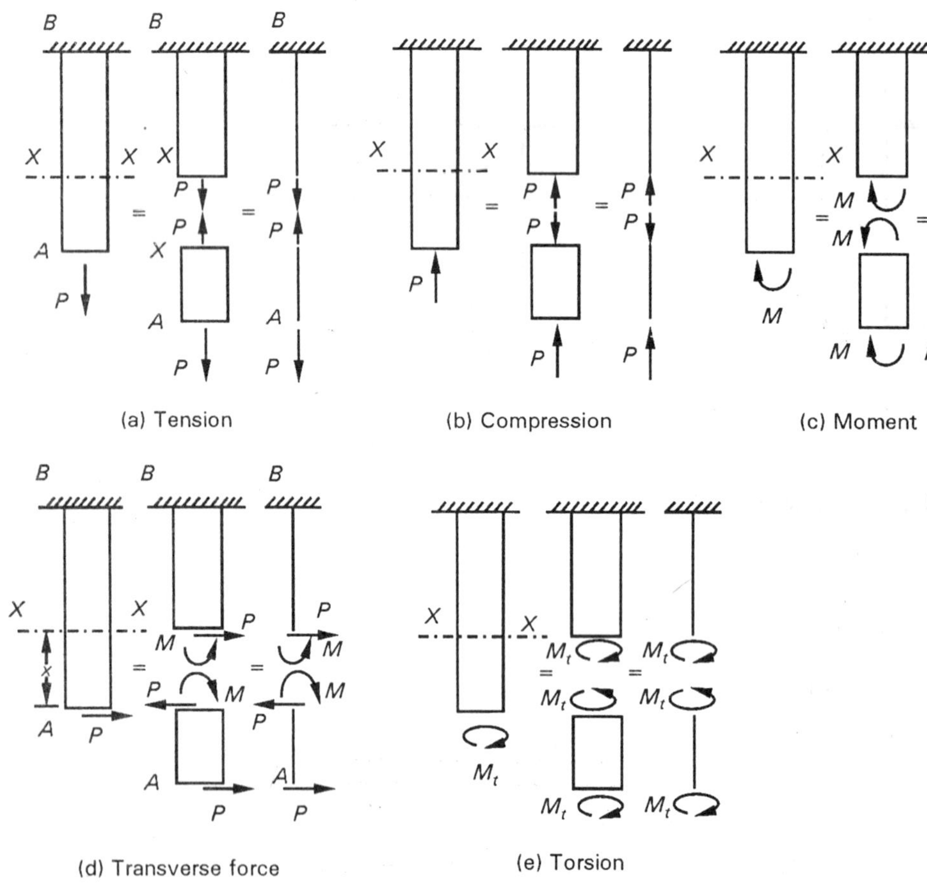

Fig.I.5 Internal forces (F.B. diagrams)

Similar conclusion can be drawn for the body of Fig.I.5(b) subjected to a compressive force, as also for the body subjected to an end moment shown in Fig.I.5(c).

However, the transverse force P acting on the body shown in Fig.I.5(d) induces a force of magnitude P to the left as well as a clockwise moment of magnitude $M(= Px)$ at section X–X of segment AX of the body. The forces at section X–X of segment XB of the body will be of the same magnitude but of opposite nature as indicated in Fig.I.5(d). In this case, the internal moment developed ($= Px$) is not constant but is a function of the distance of the section in the body from the free end; the magnitude of M varies from zero at A (free end) to Pl at B (supported end). It is obvious that the internal forces may not be constant in a body, and may vary from section to section. The determination of internal forces is necessary to compute the capacity of a body to resist external forces.

The internal forces at section X–X of a body subjected to torsional moment are indicated in Fig.I.5(e); the direction of the twisting moments on the body at various sections should be noted.

I.5 Free body diagrams

A free body diagram is a concept to visualise the internal forces of a body. An imaginary section is assumed to divide the body into segments at the required locations. The forces required to keep the body in equilibrium are noted at such sections. These segments of a body with the forces noted on the boundaries are known as *free body (F.B.) diagrams.*

Thus, a F.B. diagram illustrates a part (segment) of a structure or structural element in which the part removed is replaced by idealised forces acting at the interfaces (contact areas) of the isolated parts of the structure. The idealised forces should satisfy the equilibrium conditions of the isolated segments of the structure, however. Segments AX and XB of Fig.I.5 are F.B. diagrams by the above definition. These isolated segments are in equilibrium; the idealised forces applied at the interfaces help determine the internal stress conditions of the bodies.

I.6 Summary

The subject 'Strength of Materials' deals with material properties and behaviour of isolated members of a system. Standard conventions and symbols are indicated in this chapter. Various support conditions are explained to familiarise students with the general notation adopted in this book.

The forces acting on a body, external and internal, are explained. The concepts of F.B. diagrams are also discussed. Deformations of various types of members are indicated along with the general assumptions of the theory of analysis.

1 Direct stresses

1.0 Introduction

A typical system, such as a truss, a multi-storey building or a machine, comprises several members subjected to various types of loads. Some of the structural members may be subjected to forces along their axes. In such cases the member is said to be subjected to direct loads. Direct loading is the simplest form of loading on a member. The conditions of static equilibrium are often adequate to determine load in such members. In case the equilibrium conditions are not adequate, additional equations governing displacements have to be formulated.

The behaviour of individual members subjected to axial force (A.F.) is discussed in this chapter. Various terms of analysis are explained along with the behaviour of some typical materials subjected to axial loads.

Truss members, cables, walls and columns are some of the structural elements subjected to direct loads (tension or compression).

1.1 Stress

It is a common observation that larger sized bodies can sustain greater loads than smaller bodies of the same material. Thus, the load alone cannot be taken as the criterion to describe the capacity (or strength) of a material. Since load sustained by a body is proportional to the area over which the load is acting, the load per unit area provides a better concept of material capacity than load alone. It also helps to compare the load carrying capacities of various materials. The parameter, load per unit area is known as *intensity of stress* or, for brevity, as *stress*.

The load P acting on the body in Fig.1.1 is spread over the area A of its cross section. An A.F. transmitted across any section of a member per unit area is assumed to be uniformly distributed; the average intensity of stress (σ) can be expressed mathematically as

$$\sigma = P/A \qquad [1.1]$$

where σ = stress (positive, if tensile and negative, if compressive), P = load applied (positive, if tensile and negative, if compressive) and A = area of cross section of the body.

However, if the stress is not uniform over the section, the intensity of stress at any point (local stress) is expressed on an infinitesimal area dA over which the load dP is applied. The stress at a point is expressed in differential form as

$$\sigma = dP/dA \qquad [1.2]$$

where dA = infinitesimal area of cross section, and dP = load applied on area dA.

In elementary analysis, it is usually adequate to compute average stress given by Eqn.1.1, while Eqn.1.2 is applied in advanced analysis. The stress under direct loads is known as *direct stress*. The nature of the stress should also be specified. The stress is termed tensile or positive when the applied load causes tension (Fig.1.1(a)). The stress is compressive or negative, if the applied load causes compression (Fig.1.1(b)).

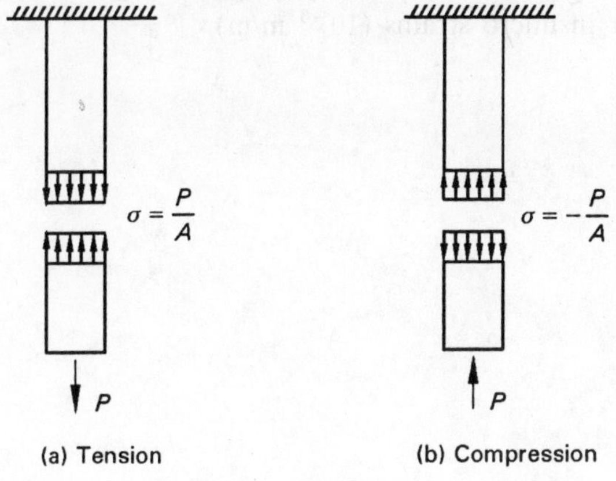

Fig.1.1 Stresses in members

1.2 Strain

The deformation of a body subjected to load is proportional to its length. A body subjected to tensile force (or stress) elongates, while compressive force (or stress) shortens it as indicated in Fig.1.2. In studying the material behaviour, the deformation per unit length of a body is of interest rather than its total deformation. The deformation per unit length is known as *strain* and is denoted by Greek symbol ε (epsilon). Mathematically, we have

$$\varepsilon = \Delta l / l \qquad [1.3]$$

where l = original length of the body, and Δl = change in the length (elongation is positive and shortening is negative).

The strain obtained from the above equation is the average value over the length l. However, if the strain is not uniform over the dimensions of a body, strain at a point (local strain) is defined over an elemental length dx. The strain at a point is defined as

$$\varepsilon = \Delta dx / dx \qquad [1.4]$$

where dx = original length of the body, and Δdx = change in the length.

It should be noted that the change in the length of a body cannot be perceived by naked eye for common materials such as steel and concrete. The small changes in length can only be measured by sensitive instruments under the usual range of loads before the material breaks or is crushed. The original length l (or dx) of the body is referred in computing strains (Eqns.1.3 and 1.4). The change in the length of a body is positive if the body elongates (positive stress); the change is negative, if the length decreases (negative stress). Accordingly, strain is positive if the body elongates (Fig.1.2(a)) and negative if it shortens (Fig 1.2(b)). The change in the dimension of a body may be due to applied loads or due to other causes such as temperature changes. The strain due to temperature changes is known as *thermal strain*. Further, strain is a ratio, and being change in length per unit length of a body, is dimensionless. However, strain is usually expressed in units of mm/mm or in micro strains (10^{-6} m/m).

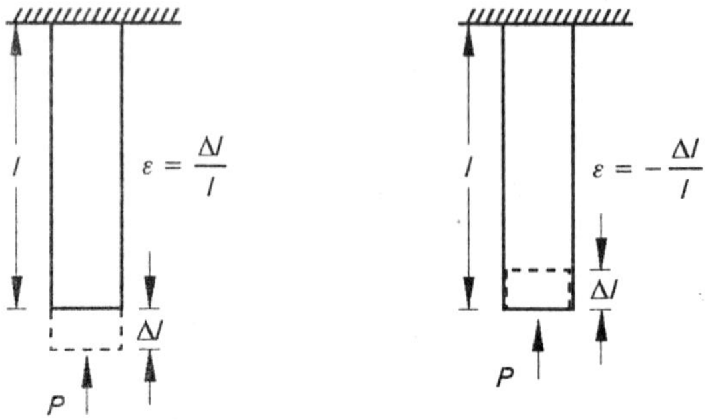

(a) Tension (b) Compression

Fig.1.2 Strains in members

1.3 Hooke's law and Young's modulus

Robert Hooke (1635–1730) investigated the behaviour of several materials subjected to gradually increasing loads. Though his methods were primitive by today's standards, he established the relationship between stress and strain in materials. Hooke's law states that *within the limit of elasticity of a material, strain is proportional to stress applied.*

A material is said to be elastic if all deformations (strains) vanish when the applied stress is removed. Hooke's law is applicable to a wide range of materials such as metals, alloys, timber and concrete. The law can be expressed as

$$\varepsilon \propto \sigma \quad \text{or} \quad \varepsilon = \sigma / E \qquad [1.5]$$

where E = proportionality constant. The constant of proportionality E (Eqn.1.5) is known as *Young's modulus* or *modulus of elasticity*; it can be expressed as

$$E = \sigma / \varepsilon. \qquad [1.6]$$

Young's modulus is thus the stress required to cause unit strain in a material. Obviously, hard materials have high values for Young's modulus; steel has a value of 190.0–210.0 GPa. Diamond is the substance with the highest value of Young's modulus (about 1 200.0 GPa), and hence least deformable under stress. Soft materials such as clay, carbon and soft stones have low values, and deform considerably more than substances with high Young's modulus (such as metals and alloys). The value of Young's modulus is usually the same in compression and tension for most of the metals and alloys. However, some materials such as concrete and clay display different characteristics in compression and tension. Unless, specifically mentioned, it is usual to assume that the value of Young's modulus is the same in compression and tension.

1.4 Stress–strain behaviour of materials

The value of Young's modulus is determined from the stress–strain graph of the material. Some materials are equally strong in compression and tension (metals and alloys); such materials are usually tested in tension. The test results usually pertain to a circular bar of uniform cross section. The load on the test specimen is increased gradually from zero, in suitable increments, till the specimen fails (breaks). The elongation of the specimen is measured over a specific length known as *gage length* (usually 50 mm to 200 mm) at each load step. The stresses and the corresponding strains are computed from the load and the corresponding elongation readings.

Materials such as concrete, stones and bricks, that are stronger in compression than in tension, are tested in compression. Stress–strain values are plotted in the form of a graph, and the value of Young's modulus is determined from the slope of the curve for any stress value. In the case of materials with linear stress–strain behaviour (Eqn.1.6), Young's modulus is constant upto the elastic limit. For materials with non-linear stress–strain relationship, the average value of the slope is adopted for Young's modulus, or the value is defined at a specified stress or strain value (for instance, at the origin, when it is known as the initial tangent modulus). The behaviour of a few typical materials is discussed in this section.

1.4.1 Mild steel

The stress–strain behaviour of mild steel is of specific interest to engineers because of the extensive use of mild steel as construction material. The behaviour is also significant because no other material has all the characteristics of mild steel. The specimen should be tested as specified in IS:1521–1972. A typical stress–strain curve of a mild steel specimen is indicated qualitatively in Fig.1.3.

A linear relationship between stress and strain can be observed initially; the linearity indicates that Hooke's law is obeyed upto a certain value of stress. The stress upto which Hooke's law is obeyed is known as σ_p, the *proportionality limit* (indicated by p in Fig.1.3). The material, however, remains elastic (that is, elongation vanishes completely on the removal of load) even beyond the limit of proportionality. The stress upto which the material remains elastic is known as σ_e, the *elastic limit* (indicated by e in the figure). If the specimen is loaded beyond its elastic limit, it will not regain its original dimensions; it suffers permanent strain known as *residual* (or *plastic*) *strain* on unloading.

If the specimen is loaded further beyond the elastic limit (point e in Fig.1.3), it reaches a point where a sudden increase in strain occurs without an appreciable increase in stress. This feature is known as yielding of the material and is represented by y in the figure; the corresponding stress is known as the *upper yield point* (σ_y). A careful testing of the specimen would reveal that the curve drops slightly at this point to y_1; the corresponding stress is known as *lower yield point* (σ_{y_1}). The specimen undergoes yielding beyond point y upto a point y_2 shown in the figure. The stress–strain curve in the region $y - y_2$ is not steady and displays waviness when tested with precision equipment. Beyond point y_2, further straining of the material can occur only with an increase in the applied load. This phenomenon is known as *strain hardening* of the material. The stress–strain curve continues to rise till u, the maximum stress the material can sustain; this value is known as the *ultimate stress* (σ_u) of the material.

It would be interesting to note that the strain between points y and u is about 100 times that between O and y. That is, the plastic strain of steel at its ultimate stress value is about 100 times its maximum elastic strain. As the specimen is stressed further, its strain (elongation) increases with a slight decrease in stress (load) accompanied by lateral contraction (decrease in bar diameter); the specimen fractures (breaks) at point f on the diagram (σ_f, *failure stress*). The specimen diameter is reduced considerably at the section of fracture; this is known as *necking*. The ends of the failed specimen form a cup and a cone as indicated in Fig.1.3.

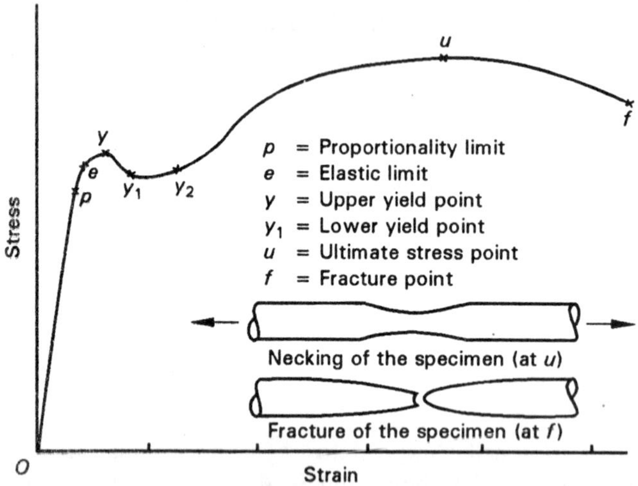

Fig.1.3 Stress–strain characteristics of mild steel

It is a general practice to compute stresses and strains with reference to the original dimensions (gage length and cross-sectional area) of the specimen. These parameters that do not consider increased length and reduction in area are known as *engineer's stress* and *engineer's strain* (or simply stress and strain), since engineers relate the parameters to original dimensions of the member. These values are also known as *nominal stress* and *nominal strain*.

It would be interesting to note that the strain at yield point of mild steel is about 1.4×10^{-3}, that at ultimate stress is about 160.0×10^{-3} and at failure it is about 350.0×10^{-3}.

1.4.2 High strength steels

The behaviour of high strength steel bars is much different from that of mild steel. The stress–strain curve of high strength steel bars is indicated qualitatively in Fig.1.4. The diagram does not indicate a well defined yield point unlike that of mild steel. The material displays linear characteristics initially (upto p). However, the curve flattens out gradually as the stress is increased beyond 80 percent of its ultimate value. The maximum stress (σ_u) the material can sustain is reached at u and the material fails at f.

The yield stress (σ_y) for such materials without a well defined yield point is determined from the criterion of *proof stress*. Proof stress is defined as the stress which induces a specified residual strain (usually 0.2 percent) in the material. This is sometimes known as *0.2 percent proof stress* or simply *proof stress*. A line parallel to the initial straight line part of the curve is drawn from the 0.2 percent strain on the X-axis to intersect the stress–strain curve at y, which yields the proof stress of the material (Fig.1.4).

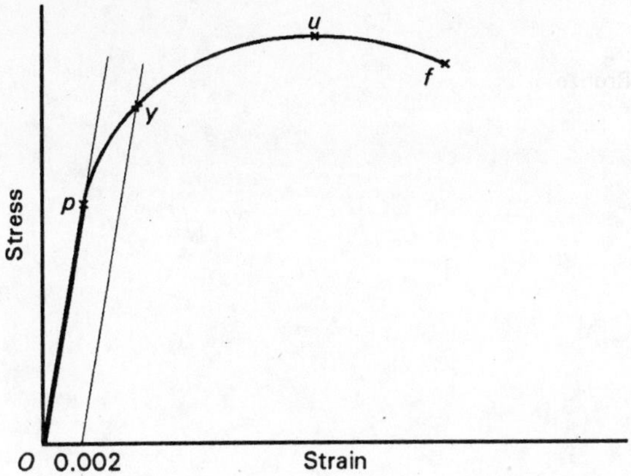

Fig.1.4 Stress–strain characteristics of high strength steels

1.4.3 Other materials

Every material has its own strength characteristics (stress and strain values at yield, ultimate and fracture stages). The stress–strain curves of a few typical materials are indicated in Fig.1.5. Students are advised to note the differences in the behaviour of various materials. Typical values of unit weights and elastic properties for a few metals and alloys are indicated in Table 1.1.

Table 1.1 *Physical and elastic properties of a few metals and alloys (gage length = 50.0 mm)*

S.No	Metal/alloy	Unit weight (kN/m^3)	Proportionality limit (MPa)	Ultimate strength (MPa)	E (GPa)	Poisson's ratio	Percentage elongation
1	Mild steel	78.5	240.0	410.0	200.0	0.30	35.0
2	HSD steel	78.5	420.0	550.0	200.0	0.30	18.0
3	Cast iron	72.0	40.0	140.0	100.0	0.25	3.0
4	Wrought iron	77.0	210.0	370.0	170.0	0.33	35.0
5	Aluminium (cast)	26.5	60.0	90.0	70.0	0.22	20.0
6	Aluminium (alloy)	27.0	220.0	390.0	70.0	0.22	25.0
7	Brass (rolled)	85.0	170.0	380.0	100.0	0.25	30.0
8	Bronze (cast)	82.0	140.0	230.0	80.0	0.14	10.0
9	Copper (hard drawn)	88.0	260.0	380.0	120.0	0.48	4.0

The above table presents average values of the parameters; these values vary depending upon the manufacturing process and method of testing. Some materials like timber have different properties in different directions (along and across the grain of timber). The direction and the nature of test (tension or compression) are also to be specified in such cases (IS:1708–1988).

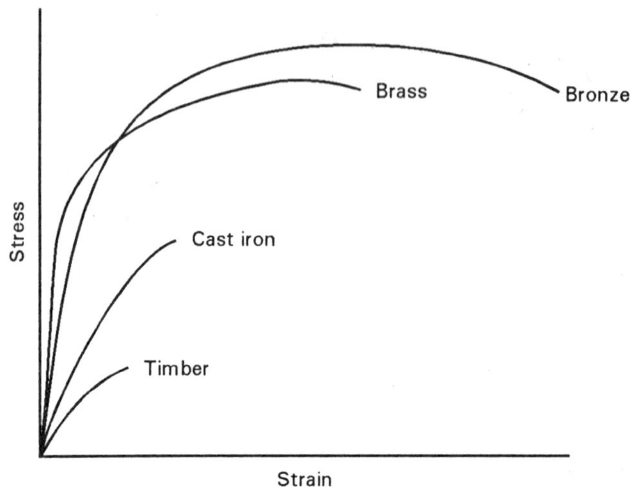

Fig.1.5 Stress–strain characteristics of a few materials

1.4.4 Ductility

Figure 1.6 indicates the stress–strain curves of mild steel, high strength steel, brass, aluminium and concrete. It can be noted from the figure that not only do the ultimate stress values differ but the strain at fracture is also different for the materials. Mild steel sustains much larger strain at failure than a HSD bar for instance. The property of a material to sustain large strains at fracture is known as *ductility*.

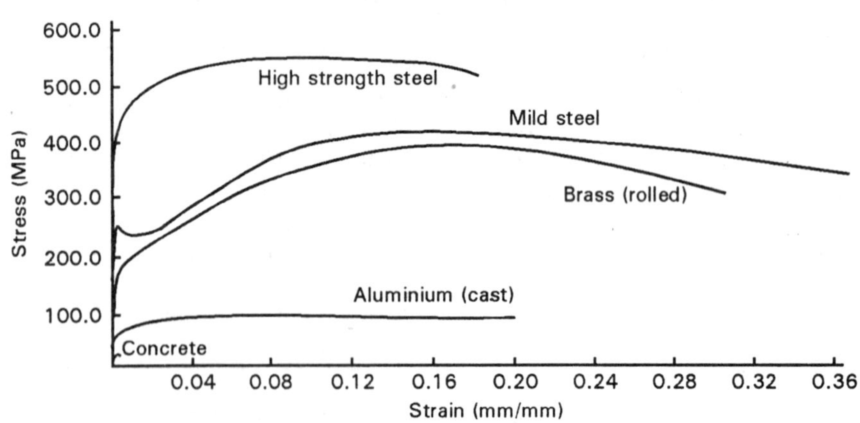

Fig.1.6 Stress–strain diagrams of a few materials (comparison)

The ductility of a material is usually expressed in terms of percentage elongation at fracture as indicated below.

$$\text{Percentage elongation} = (l_u - l)/l \quad [1.7]$$

where l = original length, and l_u = length at fracture.

The gage length over which the strains are measured should also be specified, because of high local deformations (necking) in ductile materials (Fig.1.3).

Ductility is sometimes expressed as percentage contraction of cross-sectional area at fracture as defined below.

$$\text{Percentage contraction} = (A - A_u)/A \quad [1.8]$$

where A = original area of cross section, and A_u = area of cross section at fracture.

The percentage contraction is smaller than percentage elongation for high carbon steels (carbon more than 0.3 percent).

It can be noted from Fig.1.6 that the larger the plastic range the greater is the ductility of the material. Since elastic range of a material is much smaller than its plastic range, the ductility of a material is expressed as the percentage strain (or elongation) at fracture. It indicates the ability of a material to be drawn into thin wires.

Mild steel can sustain strains upto 35 percent at fracture, while high strength steel fails at strains of about 18 percent and high carbon steels with an ultimate strength of 1 000.0 MPa fail at strains less than 10 percent. It can be noted that the strain at fracture of steels decreases as their ultimate strength increases; that is, their ductility is reduced with an increase in strength.

While the curves of all materials in Fig.1.6 pertain to tension tests, that of concrete pertains to compression test. The behaviour of metals is usually assumed to be the same in compression as well, except that the failure is assumed to occur at the yield stress value.

Materials are generally classified as brittle when the percentage elongation at fracture on a gage length of 100.0 mm is less than 5 percent. Concrete, for instance, fails at a much smaller strain (around 0.35 percent) than metals. Cast iron, stone and brick are a few other brittle materials.

Materials, such as aluminium and mild steel, with elongation at failure greater than 15% are termed ductile; materials with elongation at failure between 5–15% are said to be of *intermediate ductility*.

1.4.5 Elastic and plastic ranges

The region of the stress–strain curve extending from the origin to the proportionality limit is known as the elastic range of the material. The region of the curve extending from the proportionality limit to the point of fracture is known as the plastic range of the material.

Ductile materials have a large plastic range, while brittle materials cannot sustain much strain beyond their elastic range (Figs.1.5 and 1.6). It can be noted from Fig.1.6 that mild steel has an elastic range of about 0.12 percent, and a plastic range of about 35 percent. Other materials have much smaller plastic range than mild steel (plastic range of brass is about 30 percent, of aluminium is about 20 percent and of concrete is about 0.3 percent).

1.5 Factor of safety

The design of a structure is based on several assumptions regarding loads and materials. It is necessary to limit the actual stress to a value much smaller than the yield stress of the material because of various uncertainties of design parameters, such as loads and material characteristics.

In order to account for these uncertainties, the yield stress of a material is reduced by a factor while specifying its allowable stress; this factor is known as the *factor of safety*. The working stress σ_w (maximum permissible or allowable stress) of a material is computed by dividing its yield stress by the specified factor of safety. The actual stress in a material should not exceed its *working stress*, σ_w under any loading condition.

$$\sigma_w = \sigma_y \ / \ \text{factor of safety} \qquad [1.9]$$

The factor of safety value depends upon several parameters. It is smaller for materials of high reliability such as steel, which is a factory product, than for concrete made at site. Usually, a safety factor of about 1.8 is adopted for steel and about 3.0 for concrete.

1.6 Elongation of a bar

It follows from Eqns.1.1 and 1.6 that

$$\varepsilon = \sigma/E = P/(AE)$$

The elongation of a bar of length l is thus (Eqn.1.3)

$$\Delta l = Pl/(AE) \qquad [1.10]$$

It can be noted from the above equation that elongation of a bar is proportional to the load applied and its length, while it is inversely proportional to the area of cross section and the Young's modulus of the material.

1.7 Principle of superposition

If the proportionality between stresses and the resultant strains in a material is linear (which is usually the case), the principle of superposition is applicable.

The principle of superposition states that the effect of a set of loads on a body is the same as the sum of the effects caused by each load acting separately.

The effects mentioned above include all parameters such as stress, strain and deflection. It should be noted that the principle of superposition is not valid for materials with non-linear stress–strain characteristics (for instance, concrete and soils). However, within the range of working stress, the principle of superposition can be assumed to be valid even for some of such materials. The principle simplifies the analysis considerably, but is not valid beyond the yield stress of materials; advanced methods are applied in such cases.

Problem I A steel specimen of 12.0 mm diameter was tested in tension. The deformations were measured over a gage length of 100.0 mm. The load at the proportionality limit (P_p) was found to be 26.15 kN and the extension to be 0.118 mm. The load at yield (P_y) was 28.51 kN and the ultimate load (P_u) was 32.25 kN. The length of the specimen between the gage marks was measured to be 131.0 mm and the neck diameter to be 10.120 mm after fracture. Compute the Young's modulus, stresses at proportionality limit and, yield and ultimate loads of the material. Determine the percentage elongation and contraction.

Solution I We have

Area of cross section $A = \pi d^2/4 = \pi(12.0^2)/4 = 113.10 \text{ mm}^2$

Stress at proportionality limit $\sigma_p = P_p/A$

$= [26.15 \times 10^3/113.10] \text{ Nmm}^{-2} = 231.2 \text{ Nmm}^{-2}$

Yield stress $\sigma_y = P_y/A = [28.51 \times 10^3/113.10] \text{ Nmm}^{-2}$

$= 252.1 \text{ Nmm}^{-2}$

Ultimate stress $\sigma_u = P_u/A = [32.25 \times 10^3/113.10] \text{ Nmm}^{-2}$

$= 285.1 \text{ Nmm}^{-2}$

Strain at proportionality limit $\varepsilon_p = (\Delta l)/l$

$= 0.118/100.0 = 1.18 \times 10^{-3} \text{ mm /mm}$

Young's modulus $E = \sigma_p/\varepsilon_p = 231.2/(1.18 \times 10^{-3})$

$= 195.9 \times 10^3 \text{ Nmm}^{-2} = 195.9 \text{ GNm}^{-2}$ (or GPa)

Percentage elongation $= [(\Delta l)_u/l] \times 100$

$= [(131.0 - 100.0)/100.0] \times 100.0 = 31.0$

Area of cross section at fracture $= [\pi(10.12^2)/4] = 80.44 \text{ mm}^2$

Percentage contraction $= [(A - A_u)/A] \times 100$

$= [(113.10 - 80.44)/113.10] \times 100 = 28.9$

Problem II Determine the maximum permissible load on a steel bar of 20.0 mm diameter, when the factor of safety is 1.8 and the ultimate stress is 495.0 MPa.

Solution II We have

Area of cross section of the bar $= 314.16 \text{ mm}^2$

Ultimate load = ultimate stress × area of cross section = 495.0 × 314.16 N = 155.51 kN

Maximum permissible load = ultimate load/factor of safety = (155.51/1.8) kN = 86.39 kN

Problem III A mass of 1 500.0 kg is suspended from a 10.0 m long steel rope. Design the steel rope comprising 3 mm thick strands, if the elongation of the rope is limited to 20.0 mm. (i) Neglect the self-weight of the rope and (ii) Consider the self-weight of the rope as well.

Assume a Young's modulus of 195.0 GPa and a working stress of 500.0 MPa for the steel rope.

Solution III Design of a member comprises determining the area of its cross section required to satisfy the specified conditions. The rope should satisfy the conditions of maximum permissible stress (working stress) as well as that of maximum deformation specified (20.0 mm, in this case). The required area of cross section of the rope should be determined first. The elongation of the rope under the applied load should be computed; the area of cross section should be proportionately increased, if the elongation is more than 20.0 mm. The required number of strands to form the rope can be determined for the computed area.

Given data: $\sigma_w = 500.0 \text{ Nmm}^{-2}$ and $E = 195.0 \times 10^3 \text{ Nmm}^{-2}$

Weight of the body suspended: $W = mg = 1\ 500.0 \times 9.81 \text{ N} = 14\ 715.0 \text{ N}$

$$\text{Self-weight of the rope} = Al\rho = (A \times 10^{-6}) \times 10.0 \times 78.5 \text{ kN}$$

$$= (785.0 \times 10^{-3} A) \text{ N}$$

where A = area of cross section in mm^2, l = length of the rope in metres, $\rho = 78.5 \text{ kNm}^{-3}$ for steel (unit weight).

Area of a 3 mm strand: $a = \pi(3.0 \times 3.0)/4 = 7.07 \text{ mm}^2$

(i) Self-weight of the rope neglected Area of cross section of the rope can be computed as

$$A = W/\sigma_w = 14\ 715.0/500.0 \text{ mm}^2 = 29.43 \text{ mm}^2$$

Elongation of the rope is given by (Eqn.1.10)

$$e = Wl/(AE) = 14\ 715.0 \times 10.0 \times 10^3/(29.43 \times 195.0 \times 10^3) \text{ mm} = 25.64 \text{ mm}$$

The elongation of the rope is inversely proportional to its area of cross section. Since the elongation is more than the permissible value ($e' = 20.0$ mm), the area of cross section should be increased proportionately. The required area of cross section of the rope is given by

$$A' = [e/e']A = [25.64/20.0] \times 29.43 \text{ mm}^2 = 37.73 \text{ mm}^2$$

Number of 3 mm strands required = $A'/a = 37.73/7.07 = 5.34$ (say 6)

The rope should be fabricated with at least 6 strands of 3 mm diameter to satisfy the specified conditions.

(ii) Self-weight of the rope considered Area of cross section of the rope designed above

$$A = 6 \times 7.07 \text{ mm}^2 = 42.42 \text{ mm}^2 = 42.42 \times 10^{-6} \text{ m}^2$$

Self-weight of the rope is given by

$$W' = [785.0 \times 10^{-3} A] \text{ N} = [785.0 \times 10^{-3} \times 42.42] \text{ N} = 33.30 \text{ N}$$

It can be seen that the self-weight of the rope is negligible compared to the weight suspended. Students are advised to check for the stress and elongation of the rope for the total weight of $(W + W')$.

Problem IV A rope passing round a pulley supports a mass of 900.0 kg. Find the required number of 1.6 mm strands for the rope, and determine its elongation when it is 10.0 m long on one side of the pulley and 7.5 m long on the other. Assume Young's modulus of 120.0 GPa and a working stress of 230.0 MPa. The pulley weighs 0.5 kN.

Solution IV The configuration of the pulley is indicated in Fig.1.7.

$$\text{Total weight on the pulley } P = (900.0 \times 9.81 + 500.0) \text{ N} = 9\ 329.0 \text{ N}$$

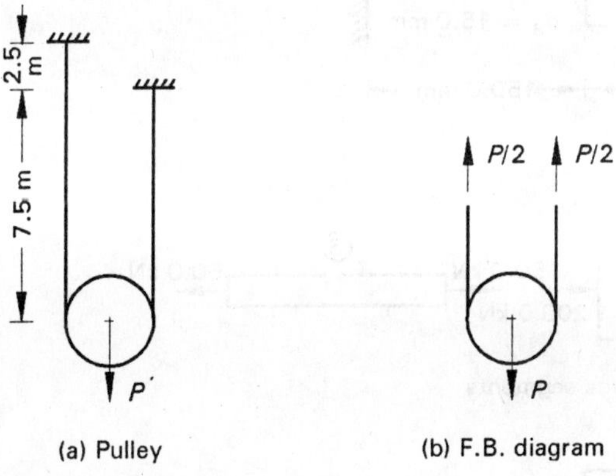

Fig.1.7 Mass suspended from a pulley (Problem IV)

The load will be shared equally by the two segments of the rope if the pulley block is frictionless (F.B. diagram of Fig.1.7(b)). In such a case, the required area of cross section A is given by

$$A = (P/2)/\sigma_w = (9\ 329.0/2)/230.0 \text{ mm}^2 = 20.28 \text{ mm}^2$$

Area of 1.6 mm diameter strand $a = 2.01 \text{ mm}^2$

Number of 1.6 mm strands required $= A/a = (20.28/2.01) = 10.1$ (say 11)

Thus, the rope should be fabricated with at least 11 strands.

Area of cross section of the rope $A = 11 \times 2.01\ \text{mm}^2 = 22.11\ \text{mm}^2$

Total length of the rope $l = 10.0 + 7.5 = 17.5$ m (neglecting the pulley)

The elongation of the rope is given by

$$e = (P/2)l/(AE) = (9\ 329.0/2) \times 17\ 500/(22.11 \times 120 \times 10^3)\ \text{mm}$$

$$= 30.8\ \text{mm}$$

Problem V Determine the stresses in various segments of the circular bar shown in Fig.1.8.

(i) Compute its total elongation assuming Young's modulus of steel to be 195.0 GPa.

(ii) Determine the length of the middle segment so that the bar length does not change under the applied loads.

(iii) Determine the diameter of segment 3 so that the bar length does not change under the applied loads.

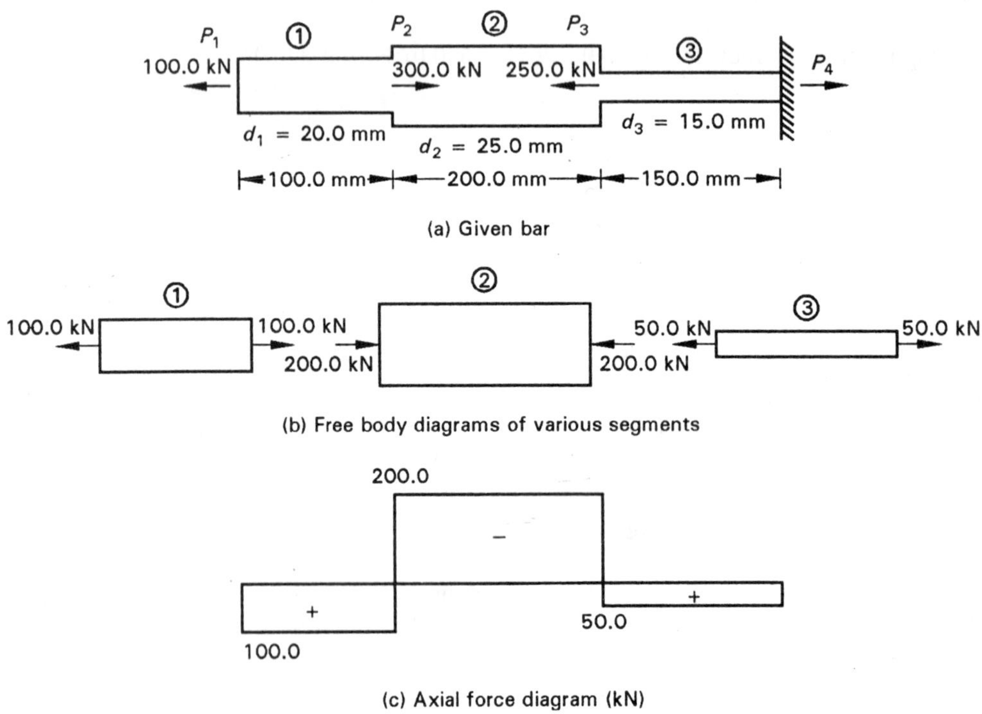

Fig.1.8 Stepped bar (Problem V)

Solution V Applying the equation of equilibrium of forces applied on the bar shown in Fig.1.8(a), we have

$$P_1 + P_2 + P_3 + P_4 = 0$$

Considering the sign convention (forces to the right are positive), we obtain

$$-100.0 + 300.0 - 250.0 + P_4 = 0$$

or $\quad P_4 = +50.0$ kN

Thus, segment 1 is subjected to a tensile force of 100.0 kN, segment 2 to a compressive force of 200.0 kN and segment 3 to a tensile force of 50.0 kN as shown in Fig.1.8(b). The stresses and deformations of the body can be determined with the help of the forces on the F.B. diagrams of the segments. Figure 1.8(c) indicates the A.F. diagram for the bar.

(i) Total elongation of the bar The computations are presented in Table 1.2 for the F.B. diagrams indicated in Fig.1.8(b).

Table 1.2 *Stress and deformation computations (Problem V)*

Segment No.	Length l (mm)	Diameter d (mm)	Area A (mm^2)	Force P (kN)	Stress $\sigma = P/A$ (MPa)	Elongation $e = \sigma l/E$ (mm)
1	100.0	20.0	314.2	+ 100.0	+ 318.3	+ 0.16
2	200.0	25.0	490.9	− 200.0	− 407.4	− 0.42
3	150.0	15.0	176.7	+ 50.0	+ 283.0	+ 0.22

Note Elongation of a bar $e = Pl/(AE) = \sigma l/E$ (Eqn.1.10);

MPa = N mm^{-2}; $E = 195.0$ GPa or 195.0×10^3 MPa

Total elongation of the bar $e = e_1 + e_2 + e_3$

$= (+0.16 - 0.42 + 0.22)$ mm $= -0.04$ mm

Negative sign indicates shortening of the bar. (Note the order of the elongations of the segments under the given loads; they are too small to be noticeable to the naked eye.)

(ii) Length of the middle segment for zero elongation of the bar The required length of the middle segment can be determined by proportioning the deformation of the segment to satisfy the given condition.

Elongation of the middle segment $= -0.42$ mm

Total elongation of the other two segments $= (0.16 + 0.22)$ mm $= +0.38$ mm

The elongation of the middle segment should be −0.38 mm so that the bar length does not change under the applied loads. Since elongation of a bar is proportional to its length, the required length l_2' of the middle segment can be proportioned as

$$l_2' = (-0.38/-0.42) \times l_2 = 181.0 \text{ mm} \qquad (l_2 = \text{original length of the segment})$$

(iii) Diameter of the third segment for zero elongation of the bar The required diameter of the third segment can be determined by proportioning the deformation of the segment to satisfy the given condition.

Elongation of the third segment = +0.22 mm

Total elongation of the other two segments = (0.16 − 0.42) mm = −0.26 mm

The elongation of the third segment should be + 0.26 mm so that the bar length does not change under the applied loads.

Since elongation of a bar is inversely proportional to its area, the required area A_3' of the segment can be computed as

$$A_3' = (0.22/0.26) \times A_3 = 149.5 \text{ mm} \qquad (A_3 = \text{original area of the segment})$$

Diameter of the third segment should be 13.8 mm (area = 149.6 mm^2) so that the total elongation of the bar is zero.

1.8 Members with varying cross section

Members analysed so far were assumed to be of uniform cross section along the length (prismatic). In practice, members may be designed with cross section varying along the length. Such members are known as *non-prismatic members*. In addition, the loading on a member may also vary along its length. In such cases, the stress distribution varies along the member length. The analysis should consider the longitudinal variation of cross section and loading along member axis.

The stresses are assumed to be constant over an infinitesimal length dx of the member (Fig.1.9). The element is at a distance x from the origin (at one end of the member). The deformation of such an element is given by

$$\Delta(dx) = \sigma_x \, dx/E$$

where σ_x = stress at a distance x from the origin.

The elongation Δl of the entire member can be obtained by integrating the above expression over its length. Thus, the member deformation is given by

$$\Delta l = \int_0^l \Delta(dx) = \int_0^l (\sigma_x/E)dx \qquad [1.11]$$

If the member is of uniform material, E can be taken outside the integration.

Problem VI A bar of uniform thickness and length l has its depth increasing from d at one end to D at the other. Determine the deformation of the member subjected to a tensile force P.

Solution VI The member is indicated in Fig.1.9. The origin is assumed at the end with smaller depth. The member depth $d(x)$ at a distance x from the origin is given by

$$d(x) = d + (D - d)x/l$$

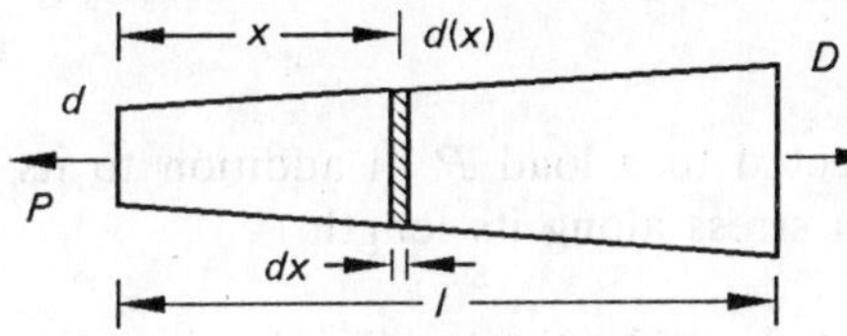

Fig.1.9 Non-prismatic bar

Let the member thickness be b. Stress at a distance x from the origin is given by

$$\sigma_x = P/[b\{d + (D - d)x/l\}]$$

Applying Eqn.1.11, the extension of the member is given by

$$\Delta l = (P/E)\int_0^l dx/[b\{d + (D - d)x/l\}]$$

$$= \left[P/(Eb)\right]\left[\ln\left\{d + (D - d)x/l\right\}_0^l\right]\left[l/(D - d)\right]$$

$$= [Pl/\{Eb(D - d)\}][\ln(D/d)]$$

For a prismatic member, $d = D$; the above expression degenerates to $\Delta l = Pl/(EA)$, where $A = bD$.

Problem VII The cross-sectional dimensions of a non-prismatic square bar vary from d at one end to D at the other. Determine the deformation of the member subjected to a tensile force P.

Solution VII Referring to Fig.1.9, the member dimension $d(x)$ at a distance x from the origin is given by

$$d(x) = d + (D - d)x/l$$

Stress at a distance x from the origin is given by

$$\sigma_x = P/[\{d + (D - d)x/l\}^2]$$

Applying Eqn.1.11, the extension of the member is given by

$$\Delta l = (P/E)\int_0^l dx/[\{d+(D-d)x/l\}^2]$$

$$= \left[P/E\right]\left[-l/(D-d)\right]\left[1/\left\{d+(D-d)x/l\right\}_0^l\right]$$

$$= [Pl/(EDd)]$$

For a prismatic member, $d = D$; the above expression degenerates to $\Delta l = Pl/(EA)$, where $A = D^2$.

Problem VIII A column of rectangular section is subjected to a load P in addition to its self-weight. Determine its longitudinal profile for uniform stress along its length.

Solution VIII Let the member be of constant thickness b. Let d be the smaller depth of the member, and D be the depth of the section at a distance x from the smaller end. Consider a segment of infinitesimal length dx as shown in Fig.1.10. The member is subjected to constant stress along its length (given). Let the member depth be $(D+dD)$ at its lower end. The forces acting on the slice are indicated in Fig.1.10(b). Assuming a unit weight ρ for the material, the equilibrium condition can be written as

$$\sigma Db + \rho Db\,dx - \sigma(D+dD)b = 0 \quad \text{or} \quad (dD/D) = \rho\,dx/\sigma$$

$$\ln(D) = \rho x/(\sigma) + C \quad \text{(from integration)}$$

where C is the constant of integration. The constant can be determined from the boundary condition that at $x = 0$, the depth of the member is d. Thus, we obtain

$$\ln(D/d) = \rho x/\sigma \quad \text{or} \quad D = d\,e^{(\rho x/\sigma)}$$

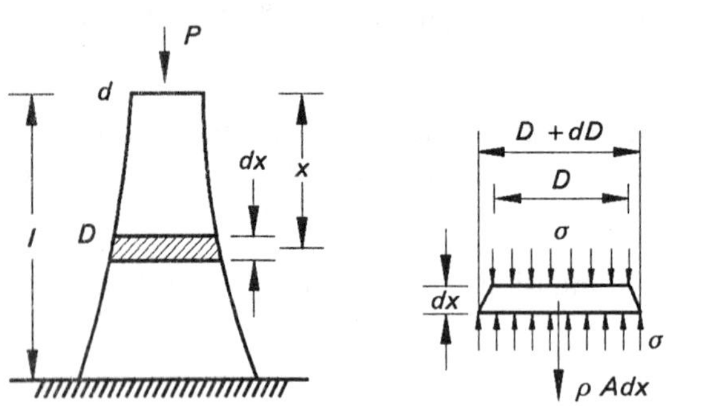

(a) Column parameters (b) Elemental parameters

Fig.1.10 Non-prismatic column with constant stress (Problem VIII)

It can be noted from the above expression that the variation of depth should be exponential for constant stress on the member. If self-weight of member is zero, the above equation yields constant depth for the member. The solution is trivial, if the applied stress σ is zero ($P = 0$).

1.9 Compound bars

Compound bars comprise two or more elements of different materials. Since more than one force (for each element) is to be determined in such cases, static equilibrium conditions alone are not adequate. Such systems, which cannot be analysed by static equilibrium conditions alone are known as *statically indeterminate* or simply *indeterminate systems*. Additional equations, using deformation conditions, are developed to supplement the equations of statics. These conditions are known as *compatibility conditions*, since the deformations in different parts of the body should be consistent and satisfy the conditions at the boundaries (compatibility means agreement).

Referring to Fig.1.11, the member comprises a bar (material 1) and a tube (material 2) assembled together. The subscripts used for forces, stresses, displacements and properties refer to the material. The force applied on the body through rigid end plates is shared by both the elements. Taking the F.B. diagram at an arbitrary section X–X (Fig.1.11), the equilibrium condition can be written as

$$P - P_1 - P_2 = 0 \qquad [1.12]$$

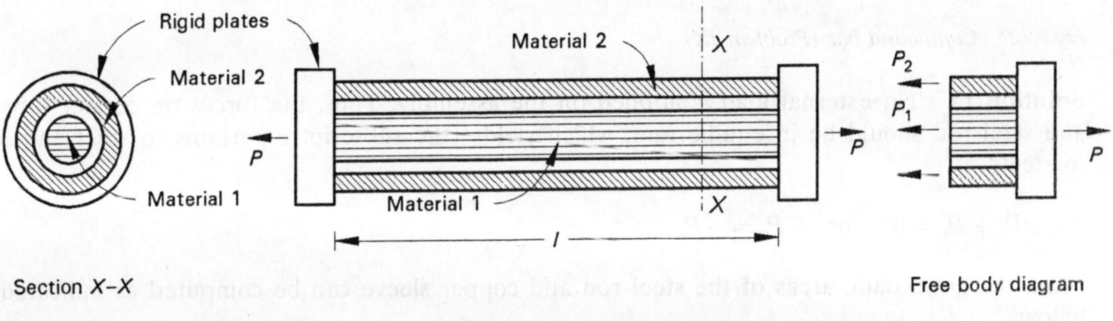

Fig.1.11 Compound bar

where P_1 = force in the bar of material 1, and P_2 = force in the tube of material 2. The above equation involves two variables, and thus cannot be solved. An additional equation can be obtained from the compatibility condition; the deformations in both the elements should be the same because the ends are held together by the assembly. The relative deformation between the elements being zero, the compatibility condition can be written as

$$\delta_1 - \delta_2 = 0 \qquad [1.13]$$

The above equation can be written in terms of the material parameters as indicated below.

$$P_1 l/(A_1 E_1) - P_2 l/(A_2 E_2) = 0 \qquad [1.14]$$

The solution of Eqns.1.12 and 1.13 (or 1.14) yields the forces in the two elements. Substituting for P_2 from Eqn.1.14 in Eqn.1.12, we obtain

$$P_1 = PA_1E_1/(A_1E_1 + A_2E_2) \tag{1.15(a)}$$

and $$P_2 = PA_2E_2/(A_1E_1 + A_2E_2) \tag{1.15(b)}$$

These problems involve some ingenuity to formulate compatibility conditions, especially in thermal problems.

Problem IX A steel rod of 15.0 mm diameter passes through a copper tube of 18.0 mm internal and 30.0 mm external diameters. The assembly is provided with nuts as indicated in Fig.1.12. The nuts are tightened to stress the steel bar to 120.0 Nmm^{-2}. Determine the stresses in copper tube, if the tube is lathe cut to remove 2.5 mm from its thickness along half of its length. Assume that Young's modulus of steel is twice that of copper. Determine the maximum compressive load the system can sustain, if the stress in copper is limited to 150.0 Nmm^{-2}.

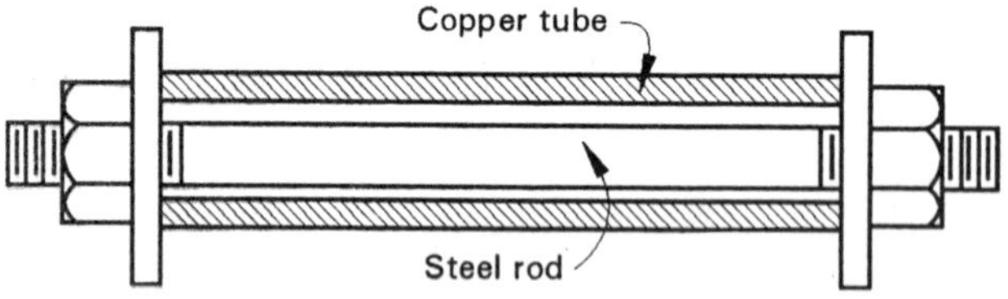

Fig.1.12 Compound bar (Problem IX)

Solution IX No external load is applied on the assembly. Thus, the forces on copper tube and steel rod should be in equilibrium; which yields (the subscript *c* pertains to steel and *s* to steel)

$$P_c + P_s = 0 \quad \text{or} \quad P_c = -P_s$$

From the given data, areas of the steel rod and copper sleeve can be computed as indicated below.

$$A_s = \pi(15.0^2)/4 = 176.71 \text{ mm}^2; \qquad A_c = \pi(30.0^2 - 18.0^2)/4 = 452.39 \text{ mm}^2$$

Since the steel rod is elongated by turning the nuts, it is subjected to tensile force (Fig.1.12). Thus, we have

$$P_s = \sigma_s A_s = 120.0 \times 176.71 = 21\ 205.20 \text{ N}$$

The stress in the copper tube is given by

$$\begin{aligned} \sigma_c &= P_c/A_c \qquad (\text{since } P_c = -P_s) \\ &= -P_s/A_c = -21\ 205.20/452.39 = -46.87 \text{ Nmm}^{-2} \end{aligned}$$

(i) Stresses due to reduced section of the sleeve If the thickness of the sleeve is reduced over half its length, the stresses in the assembly will be altered. Let the reduced area of the sleeve be A_{c1}. The outer diameter of the sleeve will now be 25.0 mm. Thus, we have

$$A_{c1} = \pi (25.0^2 - 18.0^2)/4 \text{ mm}^2 = 236.40 \text{ mm}^2$$

Let the forces and stresses in the assembly due to the reduction in the sleeve section be denoted by a prime ($'$). Applying the equilibrium condition, we have

$$P_c' = P_{c1}' = -P_s'$$

Writing the above equation in terms of stresses, we obtain

$$\sigma_c' A_c = \sigma_{c1}' A_{c1} = -\sigma_s' A_s$$

The above equations yield

$$\sigma_c' = -0.391\sigma_s' \quad \text{and} \quad \sigma_{c1}' = -0.748\sigma_s'$$

The changes in the deformations of the steel rod and the sleeve over the length l due to the forces P_c', P_{c1}' and P_s' should be equal. The compatibility condition can be written as

$$[(\sigma_s - \sigma_s')l/E_s] = [(\sigma_c - \sigma_c')l/(2E_c)] + [(\sigma_c - \sigma_{c1}')l/(2E_c)]$$

Simplifying the above equation, we obtain

$$(\sigma_s - \sigma_s') = (2\sigma_c - \sigma_c' - \sigma_{c1}') \quad (\text{since } E_s = 2E_c)$$

Substituting for σ_c' and σ_{c1}' in terms of σ_s' in the above equation, we have

$$\sigma_s - \sigma_s' = (2\sigma_c + 1.139\sigma_s')$$

Noting that $\sigma_s = 120.0 \text{ Nmm}^{-2}$ and $\sigma_c = -46.87 \text{ Nmm}^{-2}$, we obtain $\sigma_s' = 99.93 \text{ Nmm}^{-2}$. Thus, we have (final stresses in the copper tube)

$$\sigma_c' = -0.391\sigma_s' = -39.07 \text{ Nmm}^{-2}$$

and $\sigma_{c1}' = -74.74 \text{ Nmm}^{-2}$

(ii) Maximum permissible load If an external load is applied on the assembly, the load will be shared by the steel rod and the copper sleeve. The loads on the individual components can be determined by applying compatibility equation. Let the additional stresses due to P, the load applied, be denoted by double prime ($''$). Since a compressive stress of 74.74 Nmm^{-2} exists in the sleeve, the additional stress that can be permitted in the sleeve is given by

$$\sigma_{c1}'' = [-150.0 - (-74.74)] \text{ Nmm}^{-2} = -75.26 \text{ Nmm}^{-2}$$

The equation of equilibrium can be written as

$$P_c'' + P_s'' - P = 0$$

Further, we have

$$P_c'' = \sigma_c'' A_c = \sigma_{c1}'' A_{c1} \quad \text{or} \quad \sigma_c'' = 0.523\,\sigma_{c1}''$$

Applying the compatibility condition that the deformation of a steel rod should be equal to that of the sleeve, we obtain

$$[\sigma_s'' l/E_s] = [\sigma_c'' l/(2E_c)] + [\sigma_{c1}'' l/(2E_c)]$$

Simplifying the above equation, we obtain

$$\sigma_s'' = 1.523\,\sigma_{c1}''$$

Substituting for $\sigma_{c1}'' = -75.26$ Nmm^{-2}, we obtain from the equation of equilibrium (indicated above)

$$P = -38.05 \text{ kN}$$

Total stresses can be computed by adding the stress due to P to the stresses in the system.

Stress in the steel rod $= \sigma_s' + \sigma_s'' = 99.93 + 1.523(-75.26) = -14.69$ Nmm^{-2}

Stress in the copper sleeve (larger section) $= \sigma_c' + \sigma_c'' = -39.07 + 0.523(-75.26)$

$= -78.43$ Nmm^{-2}

Stress in the reduced section of copper sleeve $= \sigma_{c1}' + \sigma_{c1}'' = -74.74 - 75.26$

$= -150.0$ Nmm^{-2}

Problem X An aluminium rod of 20.0 mm diameter passes through a steel sleeve of internal diameter 22.0 mm and external diameter 33.0 mm. The sleeve is 450.0 mm long, while the rod is 0.3 mm longer than the sleeve. The assembly is supported by rigid plates at both ends and loaded in compression (Fig.1.13). Determine

(i) the load upto which only the rod is stressed

(ii) maximum load on the assembly, if allowable stresses in aluminium and steel are 120.0 and 180.0 MPa, respectively, and

(iii) the deformation of the assembly under the maximum load.

($E_s = 210.0$ GPa and $E_a = 70.0$ GPa)

Solution X Subscript a pertains to aluminium and s to steel in the computations. The assembly is shown in Fig.1.13; the rod is longer by δ (= 0.3 mm, given data) than the enclosing tube.

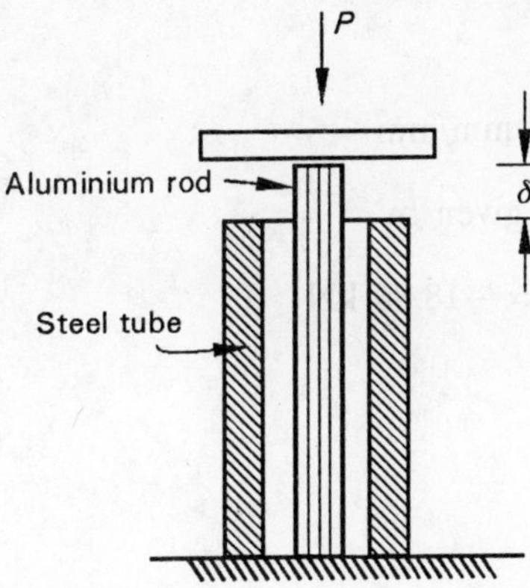

Fig.1.13 Compound bar (Problem X)

(i) Load sustained by the aluminium rod alone Since the aluminium rod is longer by 0.3 mm than the steel sleeve, the load is sustained only by the rod till its deformation is 0.3 mm. Thereafter the load will be shared by the steel sleeve as well. Load required to cause a deformation of −0.3 mm (shortening) in the rod can be computed as

$$P_a' = (-\delta/l_a)E_a A_a = (-0.3/450.3)(70.0 \times 10^3)(314.16) \text{ N} \qquad [A_a = 314.16 \text{ mm}^2]$$

$$= -14.65 \text{ kN}$$

(ii) Maximum load on the assembly The maximum load on the assembly is governed by its allowable stresses in the materials; the first material to reach its allowable stress should be determined. Let the load on the system be P; the equilibrium condition can be expressed as

$$P = P_a + P_s$$

The maximum load the aluminium rod can sustain is given by

$$P_a = \sigma_a A_a = -120.0 \times 314.16 \text{ N} = -37.70 \text{ kN}$$

The strain in the aluminum rod due to the additional load of (37.70–14.65) kN will be

$$\varepsilon_a = -(37.70 - 14.65) \times 10^3/(314.16 \times 70.0 \times 10^3) = -1.048 \times 10^{-3} \text{ mm/mm}$$

The strain in the steel sleeve due to the additional load will be the same as the additional strain in the aluminum rod (compatibility condition). The stress in the steel sleeve can thus be computed as

$$\sigma_s = \varepsilon_a E_s = (-1.048 \times 10^{-3})(210.0 \times 10^3) \text{ N} = -220.11 \text{ Nmm}^{-2} \qquad [\varepsilon_s = \varepsilon_a]$$

This value exceeds the allowable stress in steel. Thus, the assembly cannot sustain the load computed from the allowable stress condition of aluminium. The stress in steel will be critical and thus governing. Maximum load the steel sleeve can sustain can be computed as

$$P_s = \sigma_s A_s = -180.0 \times 475.17 \text{ N} = -85.53 \text{ kN} \qquad [A_s = 475.17 \text{ mm}^2]$$

The corresponding strain in the assembly is given by

$$\varepsilon_s = \varepsilon_a = -180.0/(210.0 \times 10^3) \text{ mm/mm} = -0.857 \times 10^{-3} \text{ mm/mm}$$

The load on the aluminum rod corresponding to this strain is given by

$$P_a'' = \varepsilon_a\, E_a\, A_a = (-0.857 \times 10^{-3})(70.0 \times 10^3)(314.16) \text{ N} = -18.85 \text{ kN}$$

Total load on the aluminum rod can be computed as

$$P_a = P_a' + P_a'' = (-14.65 - 18.85) \text{ kN} = -33.50 \text{ kN}$$

Maximum load on the assembly is given by

$$P = P_s + P_a = (-85.53 - 33.50) = -119.03 \text{ kN}$$

Problem XI A rigid body of mass m is suspended from three wires of equal length l and area A located symmetrically with reference to the centroid of the suspended body. Determine the forces and the deflection of the mass, if the central wire is of aluminium and the outer wires are of steel. What will be the stress in the wires, if the steel wires are of half the section of aluminium? Assume Young's modulus of steel to be twice that of aluminium.

Solution XI Parameters with subscript s refer to steel and those with subscript a pertain to aluminium. The load system is indicated in Fig.1.14.

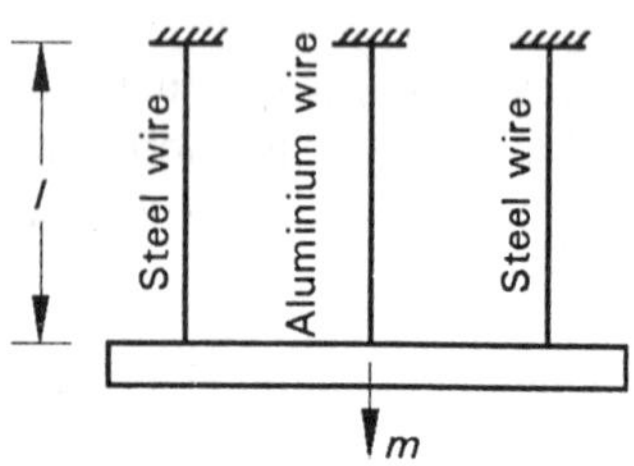

Fig.1.14 Mass suspended from three wires (Problem XI)

Given data: $E_s = 2E_a$ and $l_s = l_a = l$

Applying the condition of equilibrium, we have

$$P_a + 2P_s = m\,g \qquad \text{[XI.1]}$$

Since the load is applied symmetrically on a symmetric system of wires, the deformations of all the wires will be equal; the compatibility condition can be written as

$$P_a\, l/(A_a\, E_a) = P_s\, l/(A_s\, E_s) \qquad \text{[XI.2]}$$

(i) All wires are of the same cross-sectional area Equation XI.2 can be simplified when $A_a = A_s$ as indicated below ($E_a = E_s/2$).

$$P_a = P_s/2 \qquad \text{[XI.3]}$$

From Eqns.XI.1 and XI.3, we obtain

$$P_s = 0.4\,m\,g \quad \text{and} \quad P_a = 0.2\,m\,g$$

(ii) Cross-sectional area of aluminium wire twice that of steel wires If the cross-sectional area of the steel wires is half that of the aluminium wire ($A_s = A_a/2$), the compatibility condition (Eqn.XI.2) can be simplified as $P_a = P_s$.

The equilibrium condition (Eqn.XI.1) yields

$$P_s = P_a = (m\,g/3)$$

Note It can be noted that the stronger the member (larger E) the larger will be the load sustained by it in an indeterminate system.

Problem XII A weight of 100.0 kN is suspended from a rigid bar hanging from three wires of same material and cross section. The lengths of the wires are 8.0 m, 12.0 m and 10.0 m, respectively; the wires are spaced at 3.0 m and the weight acts at 2.0 m from the first wire. Find the forces in the wires, if the bar was horizontal initially.

Solution XII The wires are identified by their numbers shown in Fig.1.15. Since the wires are of different lengths and the weight acts unsymmetrically, the bar may not remain horizontal after the wires deform under the given weight. The deformed system is indicated in discontinuous lines in Fig.1.15.

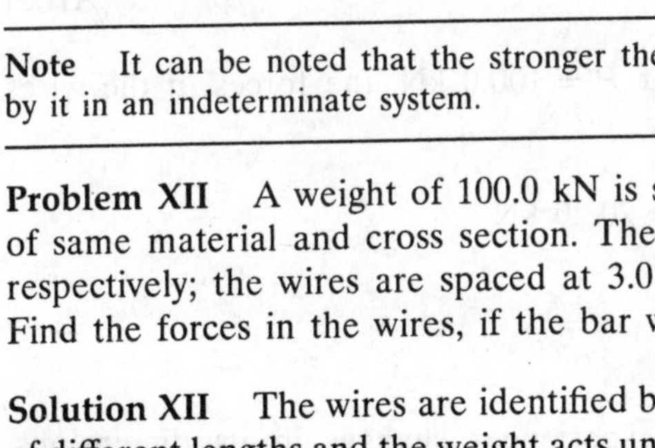

Fig.1.15 Weight suspended from three wires of unequal lengths (Problem XII)

Applying the condition of equilibrium for the vertical forces acting on the system, we have

$$P_1 + P_2 + P_3 - P = 0.0 \qquad \text{[XII.1]}$$

Taking moments of all the forces about the first wire, we obtain

$$2.0 \times P - 3.0 \times P_2 - 6.0 \times P_3 = 0.0 \qquad \text{[XII.2]}$$

Let the deformations of the wires be δ_1, δ_2 and δ_3, respectively. Since the bar is rigid, and the wires are spaced at equal distances, we have

$$\delta_2 = (\delta_1 + \delta_3)/2$$

Substituting for deformations of the wires in terms of their forces, we obtain

$$P_2 l_2 = (P_1 l_1 + P_3 l_3)/2$$

or $$24.0 \times P_2 = 8.0 \times P_1 + 10.0 \times P_3 \qquad [XII.3]$$

Solving Eqns.XII.1, XII.2 and XII.3 after noting that $P = 100.0$ kN, the forces in the wires are obtained as

$$P_1 = 53.54 \text{ kN}, \quad P_2 = 26.26 \text{ kN} \quad \text{and} \quad P_3 = 20.20 \text{ kN}$$

1.10 Thermal stresses

Any material subjected to temperature variations experiences changes in its dimensions. Deformations in a body due to temperature changes depend upon the properties of the material. The increase per unit length of a member due to unit rise in temperature is known as the *coefficient of thermal expansion* of the material, and is usually denoted by the Greek letter α (alpha). The units of coefficient of thermal expansion are mm $\text{mm}^{-1}\text{K}^{-1}$ or, sometimes, K^{-1} (change in temperature can be expressed in degrees Kelvin or Celsius). The change in the length l of a member can thus be expressed as

$$\Delta l = l\alpha\,\Delta T \qquad [1.16]$$

where ΔT = change in the body temperature. The change in temperature is considered positive, if the temperature increases (that is, the final temperature is greater than the initial temperature), and negative when the temperature decreases. The body expands when subjected to a positive change in temperature, and shortens when subjected to a negative change in temperature. By the definition of strain (Eqn.1.3), we have

$$\varepsilon_T = \alpha\,\Delta T \qquad [1.17]$$

Subscript T in the above equation indicates that the strain is due to temperature changes; such a strain is termed *thermal strain*. The convention for temperature changes is consistent with that for strains (elongation positive and shortening negative).

1.10.1 Thermal stresses in simple bars

If a body is free to deform when subjected to temperature changes, the stress in the body is zero. That is, the body is not subjected to stresses, though strain (or deformation) does exist as indicated in Fig.1.16(a). However, if the body is restrained against deformations (not free to deform) as shown in Fig.1.16(b), it develops stresses in proportion to the restrained strains. If the final strain in a body is ε, the resultant stress in a body can be computed from Eqn.1.6 as

$$\sigma_T = E(\varepsilon - \varepsilon_T) \quad [1.18]$$

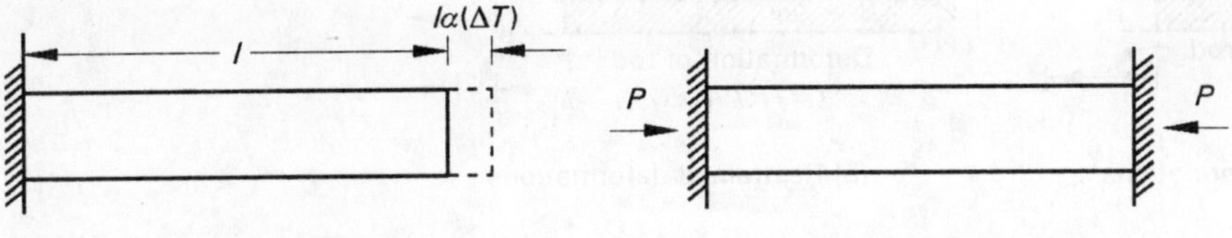

(a) Unrestrained (b) Restrained ($P = AE\varepsilon_t$)

Fig.1.16 Thermal stress in a member

In case thermal deformation is fully restrained, the final strain $\varepsilon = 0$; the thermal stress in such a member is given by

$$\sigma_T = -E\,\varepsilon_T \quad [1.19]$$

The negative sign indicates that the stress is compressive when the temperature increases (or ΔT is positive). The stress induced in a restrained member is tensile (positive) when there is a drop in temperature.

1.10.2 Thermal stresses in compound bars

Thermal stresses in compound bars can be computed from Eqn.1.19, if both the ends are held rigidly. However, unlike a simple bar, a compound bar develops stresses even when it is free to deform when the ends are held together (due to difference in the coefficients of thermal expansion of the materials). The deformation of a compound bar is illustrated in Fig.1.17. The compound bar comprises an inner rod of material 1 and an outer tube of material 2. The deformations of the tube and the rod are indicated in Fig.1.17(a), when both are free to expand; it is assumed that the material of the rod has higher coefficient of thermal expansion and elongates more than the tube. In such a case the bar elements do not develop any stresses.

However, if the ends of the bar elements are held together and are not free to expand independently, they exert restraint on one another. The system develops internal forces as a result (Eqn.1.18), and develops a deformation such that the bar elements satisfy equilibrium as well as compatibility conditions. Obviously, the rod with higher coefficient of thermal expansion will not be allowed to expand fully by the tube with a lower coefficient of thermal expansion. The final deformations of the rod and tube are indicated in Fig.1.17(b). The forces developed in the bar due to restrained thermal expansion are also indicated in the figure. The equilibrium condition can be written as

$$P_1 + P_2 = 0 \quad \text{or}$$

$$P_1 = -P_2 = -P \quad \text{(say)} \qquad [1.20(a)]$$

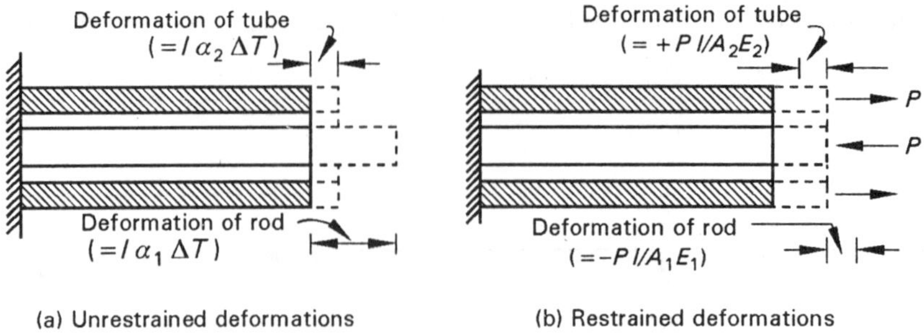

Fig.1.17 Thermal stresses in a compound bar

The rod develops a compressive force due to reduced elongation and the tube develops a tensile force due to forced elongation by the rod. The compatibility condition can be developed by considering the deformations due to temperature rise and internal forces as indicated below.

$$\delta_{1T} + \delta_1 = \delta_{2T} + \delta_2 \qquad [1.20(b)]$$

In the above equation, subscripts 1 and 2 refer to materials and T refers to temperature effects. Substituting for various deformations in the above equation, we obtain

$$[l\,\alpha_1\,\Delta T\,] + [P_1 l/(A_1E_1)] = [l\,\alpha_2\,\Delta T\,] + [P_2\,l/(A_2E_2)] \qquad [1.20(c)]$$

Substituting the equilibrium condition (Eqn.1.20(a)) in the above equation and cancelling l on both sides, we obtain

$$[\alpha_1\Delta T\,] - [P/(A_1E_1)] = [\alpha_2\Delta\,T\,] + [P/(A_2E_2)] \qquad [1.21]$$

The internal force developed in the compound bar (P) can be obtained by solving Eqn.1.21. It should be noted that in Eqn.1.21, material 1 refers to that with higher coefficient of thermal expansion.

Problem XIII The rails of a railway track are welded together at an ambient temperature of 25°C. Determine the stresses induced in the rails when the ambient temperature
(i) increases to 62°C, and
(ii) drops to −20°C.

Solution XIII The rails are assumed to be fully restrained from movements. The stresses can be computed from Eqn.1.18. Young's modulus for steel can be assumed to be 200.0 GPa and coefficient of thermal expansion as $11.0 \times 10^{-6}\ \text{K}^{-1}$.

(i) Temperature increases to + 62°C Change in the temperature of the rails

$$\Delta T = 62.0 - 25.0 = 37.0\ \text{K}$$

Thermal stress in the rails can be computed as

$$\sigma_T = -\alpha\,\Delta T\,E = -11.0 \times 10^{-6} \times 37.0 \times 200.0 \times 10^3 = -81.4 \text{ MPa}$$

Thus, increased temperature induces a compressive stress of 81.4 MPa in the rails.

(ii) Temperature drops to −20°C Change in the temperature of the rails

$$\Delta T = -20.0 - 25.0 = -45.0 \text{ K}$$

Thermal stress in the rails can be computed as

$$\sigma_T = -11.0 \times 10^{-6} \times (-45.0) \times 200.0 \times 10^3 = +99.0 \text{ MPa}$$

The drop in temperature induces a tensile stress of 99.0 MPa in the rails.

Problem XIV A wooden wheel of 1.5 m diameter is to be shrunk-fit by a steel hoop so that the hoop stress induced in the wheel is 20.0 MPa. Determine the diameter of the hoop, and the minimum temperature increase required to slip the hoop on the wheel. Assume the thermal coefficient of steel as 11.7 $\times 10^{-6}$ K^{-1} and the Young's modulus of wood as 12.0 GPa.

Solution XIV The steel hoop should be heated so that it expands and slips over the wheel. As the hoop cools, its contraction is prevented by the wheel; this induces tensile stresses in the hoop and compressive stresses in the wheel. The induced stresses are known as *hoop stresses* since they occur along the circumference of the wheel and hoop. The required temperature depends upon the stress required in the wheel and the size of the hoop.

Let the diameter of the wheel be D and the initial diameter of the hoop be $d\,(D > d)$. Let ΔT be the increase in the temperature so that the diameter of the hoop increases to D.

Initial length of the steel hoop $= \pi\, d$

Increased length of the hoop due to temperature rise

$$= \pi\, d\,(1 + \alpha\,\Delta T\,) \quad \text{[Eqn.1.17]}$$

$$= \pi D \quad \text{[circumference of the wheel]}$$

or $$D = d\,(1 + \alpha\,\Delta T\,) \quad \text{[XIV.1]}$$

Assuming that the wheel is rigid (does not deform under the stress induced by shrink-fitting), hoop stress induced in the wheel can be computed as (Eqn.1.19)

$$\sigma = -E\;\alpha\,\Delta T \quad \text{[XIV.2]}$$

From Eqn.XIV.2, the temperature change corresponding to the required strain in the hoop to induce the specified stress in the wheel can be computed as

$$\begin{aligned}\Delta T &= -\sigma/(\alpha E)\\ &= -(-20.0)/[(11.7\times10^{-6})(12.0\times10^{3})]\\ &= +142.45^\circ\text{C}\end{aligned}$$

The initial diameter of the hoop can be computed from Eqn.XIV.1.

$$\begin{aligned}d &= D/(1+\alpha\,\Delta T\,)\\ &= 1.50/[1.0+(11.7\times10^{-6})(142.45)]\ \text{m} = 1.4975\ \text{m}\end{aligned}$$

The diameter of the steel hoop should be 2.5 mm less than that of the wooden wheel.

Problem XV A steel tube of 25.0 mm external and 15.0 mm internal diameters encloses a copper rod of 12.0 mm diameter. The assembly is held rigidly at both ends at a temperature of 20°C. Compute
(i) the stresses when the temperature is raised to 120°C, and
(ii) the maximum temperature the assembly can withstand.

Assume $E_s = 210.0$ GPa, $E_c = 100.0$ GPa, $\alpha_s = 11.0\times10^{-6}\ \text{K}^{-1}$, $\alpha_c = 18.0\times10^{-6}\ \text{K}^{-1}$. Maximum permissible stress in steel is 230.0 MPa and that in copper is 120.0 MPa.

Solution XV We have

Cross-sectional area of steel sleeve $= \pi\,(25.0^2 - 15.0^2)/4 = 314.16\ \text{mm}^2$

Cross-sectional area of copper rod $= \pi\,(12.0^2)/4 = 113.10\ \text{mm}^2$

(i) Stresses in the assembly Increase in the temperature of the assembly $\Delta T = +100$ K. From Eqn.1.21, we have

$$[\alpha_c\,\Delta T\,] - [P/(A_c\,E_c)] = [\alpha_s\,\Delta T\,] + [P/(A_s\,E_s)]$$

where subscript c indicates copper element, and subscript s indicates steel sleeve ($\alpha_c > \alpha_s$). Substituting the given values in the above equation, we obtain

$$[(18.0\times10^{-6})\times100.0]-[P/(113.10\times100.0\times10^{3})] = [(11.0\times10^{-6})\times100.0]+[P/(314.16\times210.0\times10^{3})]$$

Solving the above equation, we obtain $P = 6\ 758.39$ N. The stresses in the elements of the bar can be computed as

$$\sigma_c = (P/A_c) = -59.76\ \text{Nmm}^{-2} \quad \text{and} \quad \sigma_s = +21.51\ \text{Nmm}^{-2}$$

The copper rod is subjected to compression and the steel tube to tension.

(ii) Maximum temperature in the assembly The maximum temperature the assembly can sustain is governed by the maximum stresses in the material. Assuming that the stress in the copper rod is critical, the maximum temperature can be obtained by substituting the value of the allowable stress in the compatibility condition given above. Since the copper rod and the steel tube are subjected to forces of equal but opposite magnitude, we have

$$P = -\sigma_c A_c = +\sigma_s A_s \quad \text{or} \quad \sigma_s = -0.36\ \sigma_c$$

If ΔT is the maximum permissible change in temperature, from the compatibility condition of Eqn.1.21, we have

$$\begin{aligned} (\Delta T\ )18.0 \times 10^{-6} + \sigma_c/E_c &= (\Delta T\ )11.0 \times 10^{-6} + \sigma_s/E_s \\ (\Delta T\ )(7.0 \times 10^{-6}) &= [(-\sigma_c/E_c) + (\sigma_s/E_s)] = [(-\sigma_c/E_c) + (-0.36\ \sigma_c/E_s)] \\ &= [(-120.0)\{(-1/100.0 \times 10^3) - (0.36/210.0 \times 10^3)\}] \end{aligned}$$

Solving the above equation, we obtain the maximum permissible change in temperature $\Delta T = 200.82$ K. We should investigate whether the stress in steel is critical by substituting for σ_s in the compatibility condition.

$$\begin{aligned} (\Delta T\ )(7.0 \times 10^{-6}) &= [\{(\sigma_s/0.36)/E_c\} + (\sigma_s/E_s)] \\ &= [(230.0)\{1/(0.36 \times 100.0 \times 10^3) + 1.0/(210.0 \times 10^3)\}] \end{aligned}$$

Solving the above equation, we obtain the maximum permissible change in temperature $\Delta T = 1\ 069.16$°C. The temperature difference of 1 0C9.16°C cannot be attained, since the assembly reaches the critical value in copper rod at a temperature difference of 200.82°C itself. Thus, the maximum temperature the assembly can sustain for the given data will be

$$T_{\text{max}} = (20.0 + 200.82)^\circ\text{C} = 220.82^\circ\text{C}$$

1.11 Poisson's ratio

Any material subjected to strain in one direction undergoes strain in the transverse direction as well. It has been found experimentally that axial strain is always accompanied by transverse strain with a reversed sign. That is, axial elongation is accompanied by transverse (lateral) contraction, and axial shortening is accompanied by transverse elongation. These deformations are indicated in Figs.1.18(a) and (b). The ratio of transverse to axial strains is constant within the elastic limit for a given material and is denoted by the Greek letter ν (nu). The ratio is called Poisson's ratio after the French mathematician, who determined the ratio analytically by applying molecular theory of structure of materials. The ratio can be expressed as

$$\nu = -\ [\text{lateral strain/axial strain}] = -\varepsilon_t/\varepsilon \qquad [1.22]$$

where ε_t = lateral strain.

The value of Poisson's ratio is different for different materials; the value is nearly 0.50 for rubber and paraffin, about 0.30 for steel, 0.22 for aluminum and its alloys, and 0.15 for concrete. Poisson's ratio for cork is nearly zero.

Note It would be interesting to note that human ligaments have a negative value for Poisson's ratio; when a ligament is subjected to tensile stress, it may elongate in longitudinal as well as in transverse directions (the value of ν, Poisson's ratio, varies from $+\ 0.06$ to $-\ 0.47$).

If a bar of length l and diameter d is subjected to tension, the changed length of the bar is given by (Eqn.1.3)

$$l' = l + \Delta l = l(1.0 + \varepsilon) \qquad [1.23(a)]$$

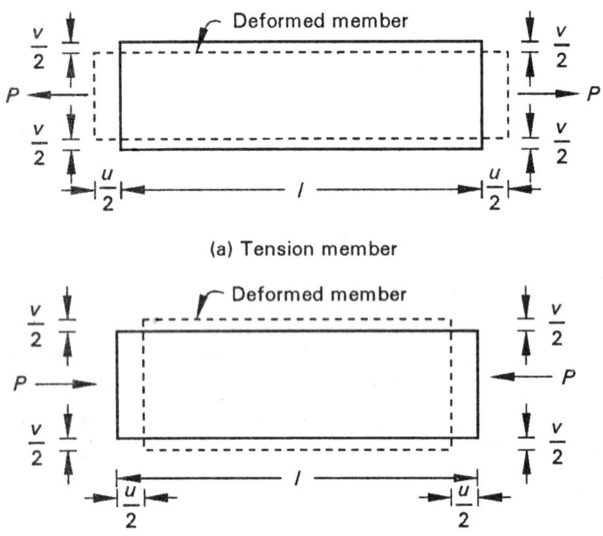

(b) Compression member

Fig.1.18 Poisson's effect (u = longitudinal deformation; v = transverse deformation)

The change in diameter is given by (Eqn.1.22)

$$\Delta d = \varepsilon_t\, d = -\nu\, \varepsilon\, d$$

The changed diameter of the bar can be expressed as

$$d' = d(1.0 - \nu\, \varepsilon) \qquad [1.23(b)]$$

The volume of the bar after loading can be expressed as

$$V' = (\pi d'^2/4)l' = \pi[d(1.0 - \nu\, \varepsilon)]^2/4]\, l(1.0 + \varepsilon)$$

The change in the volume of the bar is given by (neglecting the product of small quantities)

$$dV = V' - V = V\,\varepsilon(1.0 - 2\,\nu)$$

or $$(dV/V) = \varepsilon(1.0 - 2\,\nu) \quad [1.24]$$

That is, volumetric strain in a material subjected to axial tension is $\varepsilon(1.0 - 2\nu)$. This parameter should be a positive quantity since it is unlikely that the volume of any material subjected to tensile force will diminish in volume. Thus, the maximum possible value of Poisson's ratio should be 0.5; some materials like rubber, paraffin and clays have a value approaching this value. Similarly, axial compression will cause transverse elongation. The above expressions can be applied for compressive forces by substituting proper sign for strains and forces.

Problem XVI Determine the change in the volume of a steel bar of 25.0 mm diameter and 500.0 mm length when subjected to a tensile stress of 200.0 MPa.

Solution XVI Since it is a steel bar, $E_s = 200.0$ MPa and $\nu = 0.30$.

Axial strain in the bar $\varepsilon_s = \sigma/E_s = 200.0/(200.0 \times 10^3)$

$= 1.0 \times 10^{-3}$ mm/mm

Volume of the bar $V = \{\pi(25.0^2)/4\}(500.0)\ \text{mm}^3 = 245.44 \times 10^3\ \text{mm}^3$

Change in the volume $dV = V\,\varepsilon(1.0 - 2\,\nu)$ [Eqn.1.24]

$= 245.44 \times 10^3 \times (1.0 \times 10^{-3})(1.0 - 2 \times 0.3)\ \text{mm}^3$

$= 98.17\ \text{mm}^3$

Percentage change in volume $= (dV/V) \times 100 = [98.17/(245.44 \times 10^3)] \times 100$

$= 0.040$ [Note the magnitude]

1.12 Bulk modulus

Materials may be subjected to stresses in more than one direction. Strain–stress relations for three-dimensional state of stress can be expressed by extending Eqns.1.5 and 1.22 to three dimensions as indicated below:

$$\varepsilon_x = [\sigma_x - \nu\,(\sigma_y + \sigma_z)]/E$$

$$\varepsilon_y = [\sigma_y - \nu\,(\sigma_z + \sigma_x)]/E$$

$$\varepsilon_z = [\sigma_z - \nu\,(\sigma_x + \sigma_y)]/E \quad [1.25]$$

where σ_x, σ_y and σ_z denote stresses in X, Y and Z directions, respectively, and ε_x, ε_y and ε_z denote strains in X, Y and Z directions, respectively.

Equations 1.25 are known as the generalised form of Hooke's law. Figure 1.19 shows a body under three-dimensional state of stress. If the body has dimensions l_x, l_y and l_z in X, Y and Z directions respectively, the changed volume can be expressed as

$$\begin{aligned} V' &= l_x(1.0+\varepsilon_x)\,l_y\,(1.0+\varepsilon_y)\,l_z\,(1.0+\varepsilon_z) \\ &= l_x\,l_y\,l_z\,(1.0+\varepsilon_x+\varepsilon_y+\varepsilon_z) = V\,(1.0+\varepsilon_x+\varepsilon_y+\varepsilon_z) \end{aligned}$$

The change in volume $dV = V' - V = V\,(\varepsilon_x+\varepsilon_y+\varepsilon_z)$

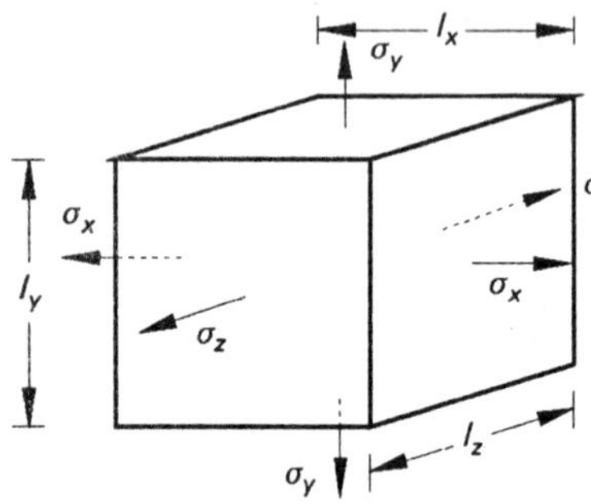

Fig.1.19 Three-dimensional stresses in a body

Volumetric strain is defined as the ratio of the change in volume to the original volume of the body. Thus, volumetric strain can be expressed as

$$\varepsilon_v = dV/V = (\varepsilon_x+\varepsilon_y+\varepsilon_z) \qquad [1.26]$$

Bulk modulus of a material is defined as the ratio of volumetric stress to volumetric strain within its elastic limit. If a body is subjected to uniform stress in all the directions, we obtain from Eqn.1.25

$$\varepsilon_x = \varepsilon_y = \varepsilon_z = \varepsilon = \sigma(1-2\nu)/E\,)$$

The bulk modulus K of the material can be expressed as

$$K = \sigma/(dV/V\,) \qquad [1.27]$$

since the stress is uniform in all the directions ($\sigma_x = \sigma_y = \sigma_z = \sigma$). Combining Eqns.1.26 and 1.27, we obtain

$$\begin{aligned} K &= \sigma/(3\,\varepsilon) \qquad [\text{since } \varepsilon = \sigma(1-2\nu)/E\,] \\ &= E/\{3(1.0-2\nu)\} \qquad [1.28] \end{aligned}$$

Equation 1.28 indicates the relationship between Young's modulus and bulk modulus of a material. The bulk modulus cannot be a negative quantity, since the volume of a material cannot increase when subjected to compressive forces or vice versa. The volume may, however, remain constant, in which case K tends to infinity (incompressible material); the corresponding value of Poisson's ratio is 0.50 — the maximum possible value for any material. It can be noted that bulk modulus of a material is a function of Poisson's ratio (Eqn.1.28); for values of ν upto 1/3, bulk modulus is smaller than Young's modulus but for values between 1/3 and 0.5, bulk modulus is larger.

Problem XVII A boiler drum of 1.0 m diameter and 3.0 m length is filled with water and subjected to a pressure of 10.0 MPa. Find the additional quantity of water to be pumped into the drum to keep it filled, if bulk modulus of water $K_w = 2.1$ GPa. Neglect the change in the volume of drum.

Solution XVII We have

$$\text{Volume of water in the drum} = \pi D^2 l/4 = 2.356 \text{ m}^3$$

$$\text{Change in the volume of water} = dV = (\sigma V/K_w) = -10.0 \times 2.356/2\,100.0 \text{ m}^3$$
$$= -0.011\,219 \text{ m}^3 = -11.219 \text{ litres}$$

Thus, additional water required to keep the drum full at a pressure of 10.0 MPa will be

$V_w = 11.219$ litres

The volume of additional water required at atmospheric pressure can be computed as

$$V_{wa} = V_w/[1.0 - (\sigma/K_w)]$$
$$= 11.219/[1.0 - (10.0/2\,100)] \text{ litres} = 11.273 \text{ litres}$$

1.13 Summary

Some structural elements, such as columns, cables and arches, are subjected to A.F. (compression or tension). The behaviour of typical structural materials subjected to A.F. is discussed in this chapter. The concepts of stresses, strains and deformations of a body subjected to forces along its axes are explained. Various parameters involved in the discussions are summarised below.

Axial stress $\sigma = P/A$ [force / area of cross section]

Linear strain $\varepsilon = \Delta l/l$ [change in length / original length]

Volumetric strain $\varepsilon_v = \Delta V/V$ [change in volume / original volume]

Young's modulus $E = \sigma/\varepsilon$ [axial stress / corresponding strain]

Bulk modulus $K = \sigma/\varepsilon_v$ [volumetric stress / corresponding strain]

Poisson's ratio $\nu = -\varepsilon_t/\varepsilon$ [lateral strain / axial strain]

Thermal deformation $\Delta l_T = l \propto (\Delta T)$ [length × coefficient of thermal expansion × change in temperature]

The significance of Hooke's law (strain in a body is proportional to the corresponding stress within the limit of elasticity), Young's modulus, Poisson's ratio and bulk modulus are also discussed. Typical problems on direct stresses are solved to illustrate the application of various concepts.

1.14 Exercises

Adopt the values of Young's modulii for various materials from Table 1.1, unless specified.

1.1 Define stress, strain and Young's modulus of a material. State Hooke's law and explain its assumptions.
1.2 Define Poisson's ratio, bulk modulus and coefficient of thermal expansion of a material. What is the maximum value of Poisson's ratio for materials? Explain with reasons.
1.3 Define proportionality limit, elasticity limit, yield stress and ultimate stress of a material.
1.4 Explain in detail the behaviour of mild steel when subjected to load test till failure.
1.5 Compare the behaviour of mild steel, high strength steel, copper, brass, aluminum and cast iron when subjected to load test till failure.
1.6 How is the yield stress of a material without a well defined yield point determined?
1.7 Define proof stress of a material and explain its significance.
1.8 Define elastic range and plastic range.
1.9 What is ductility of a material? Is ductility a desirable property of constructional material? Explain with reasons.
1.10 Define necking of a tensile test specimen, and explain the characteristics of materials in which it occurs.
1.11 What is strain hardening of a material? How does it influence the behaviour of a material?
1.12 Explain various types of stresses and deformations that can occur in a structural member subjected to A.F.
1.13 Explain the conditions of static equilibrium and their significance in analysis.
1.14 What is compatibility condition? Why is it applied in analysis? Explain with examples.
1.15 Explain the concept of factor of safety and its significance in the design of structural members.
1.16 What is working stress of a material?
1.17 Why is the ultimate strength of a material not considered in the design of a structural member? What are the usual parameters of structural design?
1.18 Which stress is considered in determining the working stress of mild steel — upper yield stress, lower yield stress or ultimate stress? Explain with reasons.
1.19 Develop an expression for volumetric strain of a body in terms of its linear strains in orthogonal directions.

1.20 What is bulk modulus of a material? Develop the relation between Young's modulus and bulk modulus of a material.

1.21 What is the influence of Poisson's ratio of a material on its bulk modulus? Discuss the behaviour of a material subjected to volumetric strains when its Poisson's ratio is (i) zero (ii) 1/ 3 and (iii) 0.5.

1.22 Describe the effects of temperature changes when a body is (i) free to deform and (ii) restrained.

1.23 Develop the compatibility condition when a compound bar of two different materials is subjected to temperature changes.

1.24 State the principle of superposition, and explain its significance.

1.25 A steel bar of 25 mm diameter was tested in tension and the following results were recorded:
Limit of proportionality = 196.32 kN, load at yield = 218.13 kN, ultimate load = 278.20 kN.
The elongation measured over a gage length of 100.0 mm was 0.189 mm at proportionality limit, length of the bar between gage marks after fracture was 112.62 mm and the minimum diameter was 23.64 mm.
Compute the stresses in the specimen at various stages, Young's modulus and percentage elongation and contraction. Determine the permissible stress in the material for a safety factor of 1.85.

1.26 A 3.5 m long steel column of cross-sectional area 5 000.0 mm^2 is subjected to a load of 1.6 MN. Determine the safety factor for the column, if the yield stress of steel is 550.0 MPa. Determine the allowable load on the column, if the deformation of the column should not exceed 5.0 mm. Assume Young's modulus for steel as 195.0 GPa.

1.27 A square concrete pedestal is to sustain a load of 500.0 kN. Determine a suitable cross section for the pedestal, if the allowable stress is 15.0 MPa. What is the allowable load, if the deformation of the 2.0 m high pedestal should not to exceed 1.0 mm? Assume Young's modulus for concrete to be 25.0 GPa.

1.28 A 2.0 m long steel tie bar is subjected to a force of 150.0 kN. Determine its cross section so that
(i) the stress does not exceed 140.0 MPa, and (ii) the extension is not more than 1.2 mm. Assume a Young's modulus of 210.0 GPa; steel bars are available in increments of 5 mm from 30 mm diameter onwards.

1.29 Compute the stresses and deformations in the steel rope of a pulley supporting a load of 126.0 kN. The rope is made of 12 wires of 2.5 mm diameter. The rope is 12.6 m on one side of the pulley and 8.1 m on the other side. Determine the elongation and the factor of safety of the system, if the yield stress for the wires is 1 800.0 MPa and the Young's modulus for the rope is 150.0 GPa.

1.30 Determine the maximum load that can be supported by a 10.0 m long steel rope comprising 12 wires of 2.0 mm diameter, if the working stress is 230.0 MPa and Young's modulus is 180.0 GPa. The rope should not elongate more than 10.0 mm under the load.

1.31 A rope comprising three 3.0 mm diameter wires clamped together at the ends, lifts a load of 15.0 kN. The lengths of the wires are 10.0 m, 10.01 m and 10.02 m. Compute the stresses in each wire. What is the maximum load the rope can lift, if the working stress is limited to 800.0 MPa and the deformation of the rope to 40.0 mm? Assume E = 210.0 GPa.

1.32 Design a steel rod to sustain a load of 80.0 kN with a safety factor of 2.5. What is the maximum permissible length of the rod, if the allowable deformation is 0.5 mm? Assume a yield stress of 230.0 MPa and Young's modulus of 195.0 GPa.

1.33 A 12.5 m long steel rope supports a mass of 1 000.0 kg. Design the rope comprising 1.6 mm strands with a yield stress of 1 500.0 MPa and Young's modulus of 175.0 GPa. Assume a safety factor of 3.5; the elongation of the rope should not exceed 12.5 mm.

1.34 Determine the distribution of stresses at salient sections of the circular bars indicated in Fig.E 1.1. Compute total deformations of the members as well.

1.35 What should be the length of the steel component in the compound bars of Figs.E 1.1(b), (c) and (d) so that the total deformation of the member is zero?

1.36 Determine the diameter of the copper component in Fig.E 1.1(b) so that total elongation of the body is zero. Relocate the 180.0 kN load such that the total elongation of the body is zero. [*Hint*: Assume the load at a distance x from the free end in the steel component. Develop the equation for total deformation and solve for x.]

1.37 Compute the diameter of the aluminum component in Fig.E 1.1(d) so that elongation of the body is zero. Relocate the 50.0 kN so that the total elongation of the body is zero.

1.38 Determine the deformation at the point of application of the load in Figs.E 1.1(e) and (f).

1.39 A bar of length l has its diameter increasing from d at one end to D at the other. Determine the deformation of the member subjected to a tensile force P.

1.40 A prismatic member of length l and unit weight w is suspended freely from its end. Determine the elongation of the member under gravity.

1.41 Determine the profile of the member of the previous problem for uniform stress distribution along its length when a load P acts at its lower end. What is the corresponding elongation of the member?

1.42 The bodies indicated in Fig.E 1.2 are rigidly held between their ends. Assuming $\alpha = 12.5 \times 10^{-6}$ and $E = 150.0$ GPa, determine the forces in the bodies when subjected to (i) a rise in temperature of 85 K, and (ii) a fall in temperature of 65 K.

1.43 Determine the forces in the systems shown in Fig.E 1.3, when the aluminum wire is subjected to a temperature fall of 50.0 K. Assume $\alpha = 23.1 \times 10^{-6}$ K^{-1} and $E = 70.0$ GPa for aluminum, and $E = 200.0$ GPa for steel; the bars are of 5.0 mm diameter. [*Hint*: Take moments of the forces about the fulcrum.]

1.44 Determine the forces in the wires of the suspended systems shown in Fig.E 1.4, if all the wires are of the same diameter and material.

1.45 Determine the forces in the wires of the suspended systems shown in Fig.E 1.5, if all the wires are of the same diameter. Assume Young's modulii of inner wires to be 1.5 times that of the outer wires.

1.46 What will be the forces in the systems of Fig.E 1.5, if the diameter of the middle wire is half that of the outer wires?

1.47 Compute the maximum force a 200.0 mm long compound bar comprising a copper rod of 18.0 mm diameter enclosed in a mild steel tube of 20.0 mm inner and 32.0 mm outer diameters can sustain. Assume Young's modulii for copper and steel to be 120.0 GPa and 195.0 GPa, respectively. What will be the reduction in the maximum load, if the bar temperature rises by 40.0 K ? Find the reduction in strength when the temperature falls by 40.0 K. Allowable stresses are 80.0 MPa and 140.0 MPa in copper and steel, respectively; $\alpha = 18.0 \times 10^{-6}$ K^{-1} for copper and 12.0×10^{-6} K^{-1} for steel.

1.48 A compound bar comprises a 12.5 mm diameter aluminum rod and a copper tube of 14.5 mm inner and 25.0 mm outer diameters. If the Young's modulii of aluminium and copper are 80.0 GPa and 120.0 GPa, respectively, determine the stresses in the assembly when subjected to
(i) a temperature rise of 95.0 K, and (ii) a temperature fall of 35.0 K; $\alpha = 14.6 \times 10^{-6}$ K^{-1} for aluminium and 16.8×10^{-6} K^{-1} for copper.

1.49 Compute the stresses in the bars of Fig.E 1.6 when subjected to a temperature (i) rise of 120 K and (ii) fall of 52 K.
Assume $\alpha = 16.8 \times 10^{-6}$ K^{-1} and $E = 120.0$ GPa for copper, and $\alpha = 11.5 \times 10^{-6}$ K^{-1} and $E = 210.0$ GPa for steel.

1.50 Determine the forces in the bars of the systems shown in Fig.E 1.7, if all the bars are of 20.0 mm diameter. Assume Young's modulii of inner bar to be 1.8 times that of the outer bars and $E = 80.0$ MPa for outer bars.

1.51 Determine the differential length between the inner and outer bars of the system shown in Fig.E 1.7(b) so that the three bars share the load equally.

1.52 What will be the forces in the systems of Fig.E 1.7, if the diameter of the middle bar is 10.0 mm, and the other conditions remain the same?

1.53 Determine the change in the volume of the bars shown in Figs.E 1.1(e) and (f). Assume a Poisson's ratio of 0.22 for aluminum and 0.3 for steel.

1.54 A boiler of diameter 2.5 m and length 4.5 m operates at a pressure of 12.5 MPa. What is the volume of water required to maintain the boiler filled at its operating pressure, if the bulk modulus of water is 2 000.0 MPa. Neglect the volume change of the boiler.

1.55 Determine the shear and bulk modulii of steel, aluminum, copper, brass and bronze for the properties indicated in Table 1.1.

1.56 A steel plate of 20.0 mm thickness is subjected to a hydrostatic pressure of 100.0 MPa. Find the change in its volume per unit surface area; assume $E = 210.0$ GPa and a Poisson's ratio of 0.28.

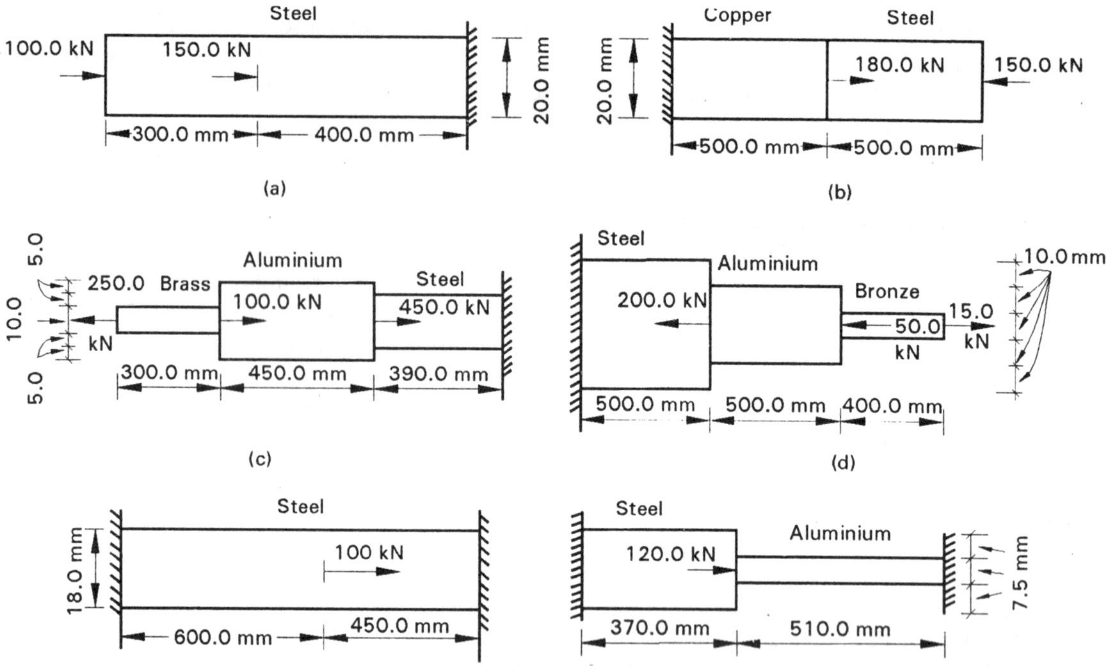

Fig. E 1.1

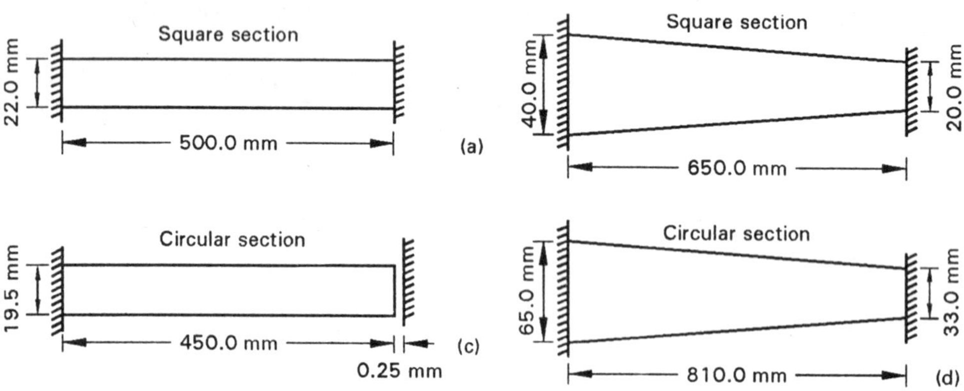

Fig. E 1.2

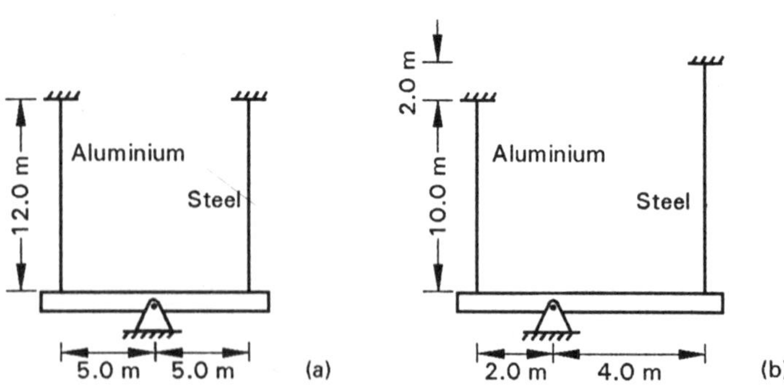

Fig. E 1.3

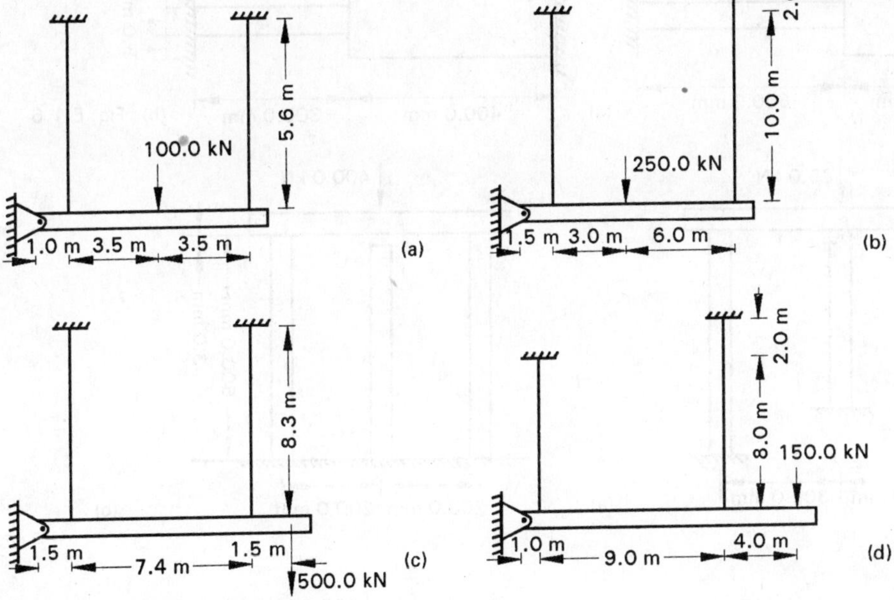

Fig E 1.4

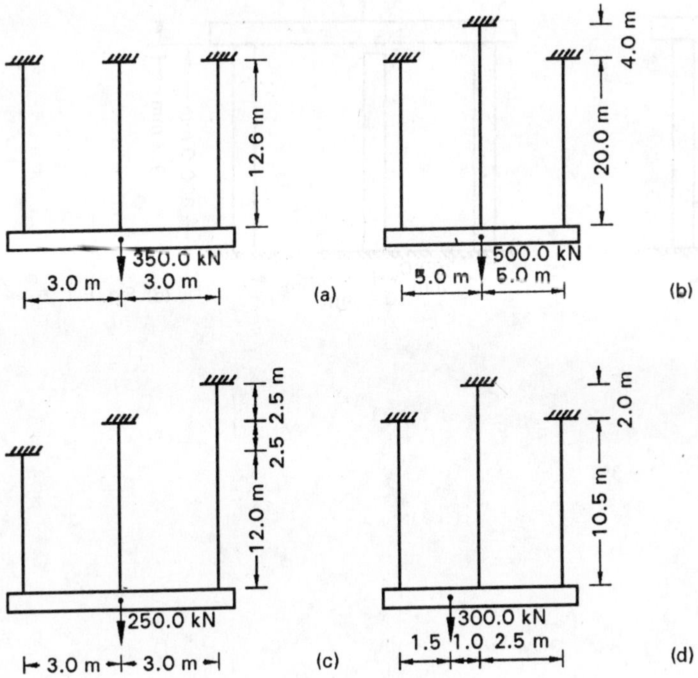

Fig. E 1.5

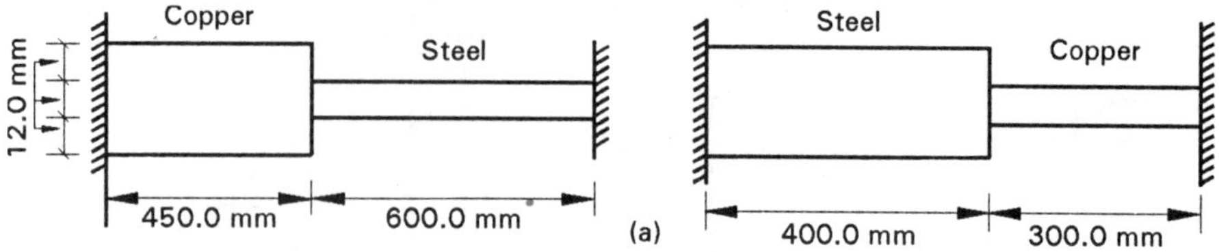

Fig. E 1.6

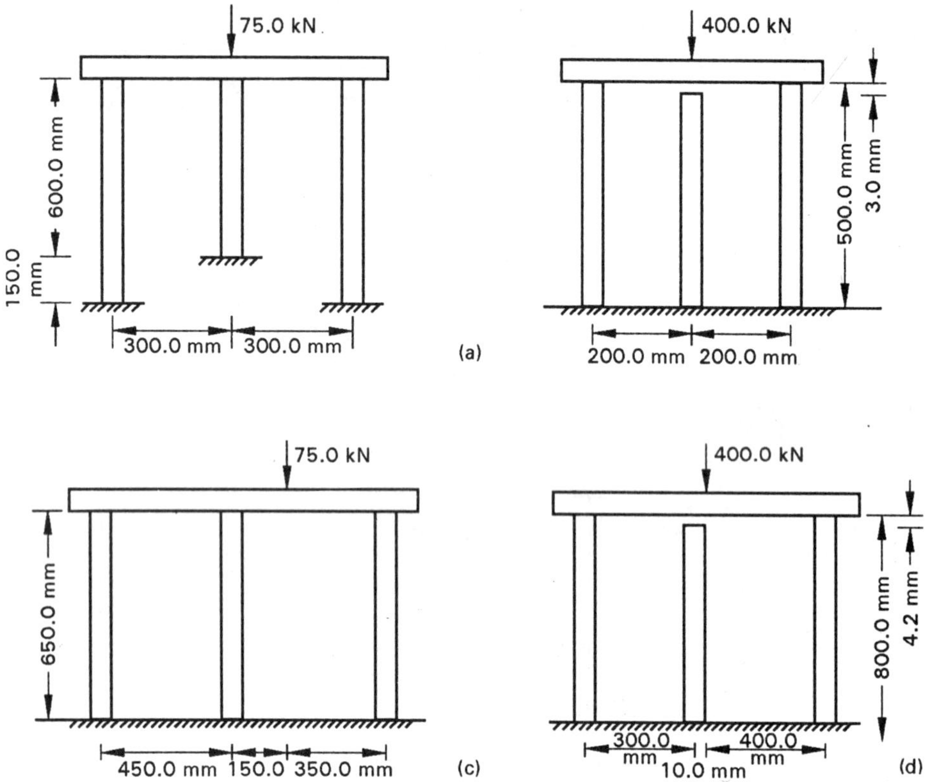

Fig. E 1.7

2 Impact loading and strain energy

2.0 Introduction

It was assumed in the discussions of Chapter 1 that loads were applied gradually on the bodies. The load (or stress) increased from zero to a specified value over a period of time. It is possible that loads are applied suddenly or with an impact (falling loads). Machine elements, for instance, are subjected to impact loads due to their movements. The effects of impact (sudden) loading on a body can be determined from energy principles. The concepts of energy along with the application to impact loading are discussed in this chapter.

2.1 Impact loading

It should be noted that unless it is specifically mentioned, loads are assumed to be applied gradually on a structure. Sudden loads imply the load increasing to its full value in an extremely short interval of time (fraction of a second). Impact loads on the other hand fall through a height on a body, and thus have additional energy.

Figure 2.1 indicates a load falling through a height h on a structural member, and causing a deflection in the member. The load falling on the collar attached to a tension member in the figure causes elongation, while that on a compression member induces shortening. The stresses increase when a load is applied suddenly or with an impact as indicated in the figure.

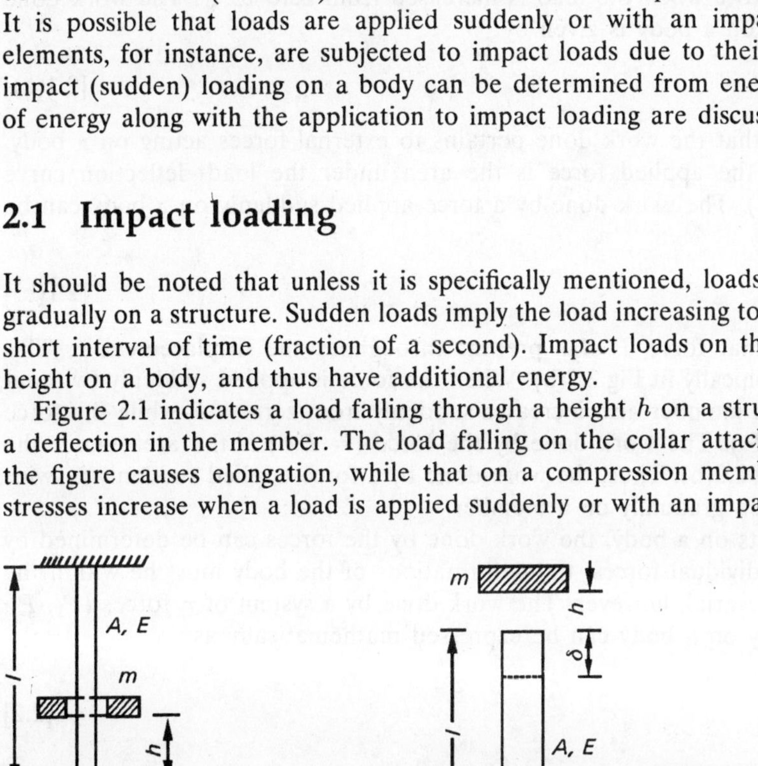

Fig.2.1 Impact loading

2.2 Work and energy of a system

The work done by a force is the product of the force and the distance of its movement. The movement may be due to the displacement of the body on which the force acts; the displacement of the body includes its deformation as well. Since the work is done by external forces acting on a body, the work is referred to as *external work*. A body develops internal forces to resist any external forces acting on it. Work is also done by the internal forces due to deformation of the body; this is known as the *internal work* or *strain energy*. Usually work is expressed in kNm or MNm, the same units as those of moment.

2.2.1 External work

Generally, a force is applied gradually on a body, increasing from zero to P, so that the displacement of the body increases from zero to δ; this condition is indicated graphically in Fig.2.2(a). Initially, at the origin, both force and deflection are zero; the deflection of the member increases gradually to δ when the load is increased from zero to P. The work done by a force applied gradually on a body is given by

$$W_e = (1/2)P\,\delta \qquad [2.1(a)]$$

The subscript e indicates that the work done pertains to external forces acting on a body. Further, the work done by the applied force is the area under the load–deflection curve (shown hatched in the figure). The work done by a force applied suddenly on a body can be expressed mathematically as

$$W_e = P\,\delta \qquad [2.1(b)]$$

Equation 2.1(b) implies that force P was present throughout the displacement δ. This condition is represented graphically in Fig.2.2(b). When the force is applied suddenly (without a drop), the deflection of the member also increases suddenly from zero to δ, while the force remains constant (equal to P). The work done by the force (= $P\,\delta$) is the area under the curve (shown hatched) in Fig.2.2(b). Thus, the work done by a force applied suddenly is twice that of the same force applied gradually on a member.

If more than one force acts on a body, the work done by the forces can be determined by superposing the effects of individual forces; the deformations of the body must be within the proportionality limit of its material, however. The work done by a system of n forces (P_1, P_2, P_3, ..., P_n) applied gradually on a body can be expressed mathematically as

$$W_e = (1/2)\sum_{j=1}^{n} P_j\,\delta_j \qquad [2.2]$$

where δ_j = displacement at the point of application of force P_j.

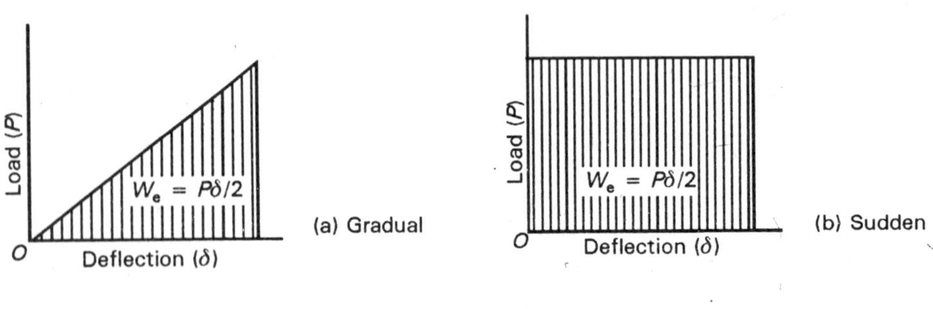

Fig.2.2 Strain energy in various materials

2.2.2 Internal work

The work done by an internal force can also be expressed in the same way as the external work; mathematically, we have

$$W_i = (1/2)P_i\,\delta_i \qquad [2.3]$$

The subscript i refers to the internal parameters of a body. The internal force can be expressed in terms of stress, and the displacement in terms of strain. Equation 2.3 can be written as

$$\begin{aligned} W_i &= (1/2)[\sigma A][\varepsilon\, l] \\ &= \{(1/2)[\sigma\,\varepsilon]\}V \qquad [2.4(a)] \end{aligned}$$

where σ = stress in the body, ε = strain in the body, l = length of the body, A = area of cross section of the body and V = volume of the body (= $A\,l$).

The internal work is also known as *strain energy* of the body. The value within the braces of Eqn.2.4(a) denotes the strain energy per unit volume of the body. Mathematically, the area of the stress–strain diagram, upto the given stress value, represents the strain energy (or internal work done) per unit volume of material. Strain energy per unit volume of a material is usually expressed in kNm/m^3 or MNm/m^3.

It is a general convention to express strain energy by the symbol U (upper case) rather than as W_i, and the strain energy per unit volume by the symbol u (lower case). Strain energy can be expressed in different forms as indicated below.

$$U = P^2 l/(2AE) \qquad [2.4(b)]$$

$$U = \sigma^2 A\,l/(2E) \qquad [2.4(c)]$$

$$U = \varepsilon^2 E A\,l/2 \qquad [2.4(d)]$$

However, the stress and strain in a body are not constant for a general case, and vary from point to point. Thus, the internal work can be expressed by considering an infinitesimal element of volume dv in the body, and integrating over its entire volume V. Equation 2.4 can be written for an elemental volume as

$$U = (1/2)\int_V [\sigma\,\varepsilon]\,dv \qquad [2.5]$$

If the stress, and hence the strain, within a body is constant, Eqn.2.5 degenerates to Eqn.2.4(a). It should also be noted that strain energy is always a positive quantity, since both stress and strain are of the same sign (positive or negative) for a given load system on a body.

The strain energy per unit volume of a body subjected to a gradually applied uniform stress equal to the proportionality limit of the material is known as the *resilience modulus*. (Resilience is the ability of a material to regain its original shape upon the removal of the applied loads.) Resilience modulus of a material is equal to the area under the corresponding stress–strain curve upto the proportionality limit from the origin. Mathematically, strain energy per unit volume can be expressed as (Eqn.2.4(c))

$$u = [\sigma^2/(2E)] \qquad [2.6(a)]$$

By definition, resilience modulus can be expressed as

$$u_r = [\sigma_p^2/(2E)] \qquad [2.6(b)]$$

It can be seen from the above equations that the larger the area under the stress–strain curve, the greater is its resilience modulus, and the greater is its ability to regain its original shape upon the removal of the applied loads. Resilience modulus of a material is a measure of the ability of the material to absorb energy within its proportionality limit.

The work done on a unit volume of a material when a force causing simple tension is increased gradually from zero upto a value causing its failure is known as its *toughness modulus*. The toughness modulus of a material is given by the area of its stress–strain curve from the origin upto fracture; it is a measure of the ability of the material to absorb energy before failure. The areas under the stress–strain curves upto fracture shown in Fig.1.6 indicate the toughness modulus of various materials; mild steel has the largest value for toughness modulus, followed by high strength steel. Concrete has very low toughness modulus compared to metals. The greater the ductility of a material, the greater is its toughness modulus and its ability to absorb energy before failure.

Materials with high resilience and toughness modulii are chosen when impact resistance is a design criterion. Structures such as bridges, machinery, motor vehicles, aircrafts and spacecrafts, require materials with high impact resistance properties.

Note It is easy to see that a small deformation of a body beyond A in Fig.2.3 results in a much larger increase in strain energy than the same deformation applied at the origin. Besides, the force required to cause the same deformation beyond A in the figure is much smaller than that required in the vicinity of the origin. This concept of strain energy is applied in everyday life. The tightened strings of a tennis racket, for instance, store large strain energy. The strain energy of the strings increases further due to the deformation caused by the impact of the ball; this energy is released on the rebound of the ball causing the ball to be hurled at great speed. It can be noticed that the players keep adjusting the strings to keep them taut. Similarly, the string of a bow is tightened to store large strain energy in the system. As the string is pulled with the shaft, the energy of the system is increased further; the increased energy is released and imparted to the shaft as kinetic energy, hurling it over a great distance. The stiffer the bow (large values of E, A and l) and the more the tension in the string, the greater will be the energy imparted to the shaft. A cupid's bow is a very efficient system; it stores large strain energy because of its initial reversed profile. A small cupid's bow can hurl a shaft over a large distance with the use of a small force as a result.

2.2.3 Energy principle

From the principle of conservation of energy, the total energy of a system remains constant.

In other words, the work done by external forces acting on a body is equal to the internal work done (strain energy). Mathematically, we have

$$W_e = W_i \tag{2.7}$$

The work done by an external system of forces during the deformation of a body is stored in the body as strain energy. The strain energy is released when the external loads are removed; that is, the body regains its original dimensions if the stresses are within the elastic limit of the material. Beyond the elastic range of a material, a part of the work done by external forces is spent in causing plastic deformations in the body. The part of work done by the external forces in causing plastic (permanent) deformations cannot be recovered upon the removal of the forces.

Figures 2.3(a) and (b) indicate typical stress–strain curves of ductile and brittle materials, respectively. If the work done by an applied load is less than or equal to the area under the stress–strain curve upto the proportionality limit of the material, the body regains its original shape. If the work done by the applied force exceeds the area under the curve upto its proporionality limit, the body suffers permanent deformation (say, area OABC). However, if the work done by external forces is greater than the area under the curve upto fracture (area OAFG), the material fractures. Figure 2.3(c) indicates the idealised bi-linear curve for computational purposes; the idealisation simplifies calculations considerably but underestimates the toughness modulus of the material and is thus conservative (errs on the safer side).

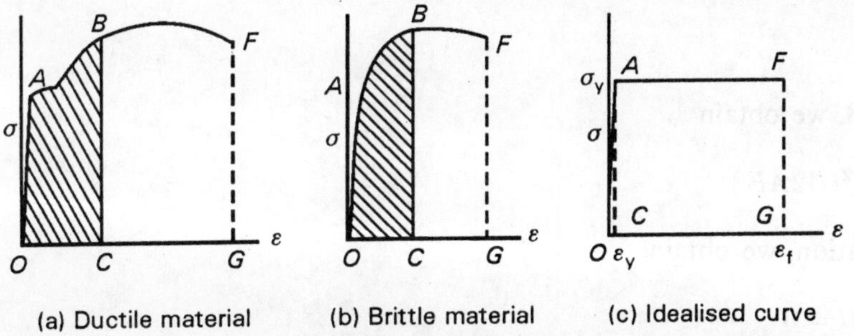

Fig.2.3 Strain energy in various materials

It is easy to comprehend that any reduction in the area of the curve (OAFG) reduces the capacity of the body to resist external loads, particularly when the loads are applied suddenly or with impact. A specimen with a small groove (reduction in its cross-sectional area) will fracture at a much smaller impact load than its capacity. This is explained in the next section through an example (Problem III).

2.3 Impact load analysis

The work done by a load when applied suddenly is more than that done by the load applied gradually (Fig.2.2 and Eqn.2.1). Further, the work done by a load dropped on a body (impact load) is much more than that by the same load applied suddenly; the work done in the former case includes the work done during the drop as well as that during the deformation of the body.

Since the external work done by a load is equal to the internal work (strain energy), it is obvious that a load causes larger deformations and stresses in a body when applied suddenly (or with impact) than when applied gradually. The member will obviously fail if the stress developed is higher than the ultimate stress of the material. However, if the stress in the body is less than its ultimate value, the body oscillates for a while and reaches its equilibrium condition. The final stresses and deformations (strains) will be the same as though the load were applied gradually. When a load is applied gradually, there is no loss of energy in the system due to such oscillations. Stresses under impact loads can be computed from the energy principles discussed in Section 2.2.

Let a load P fall through a height h on a member (bar or column) of length l, cross-sectional area A and Young's modulus E. Let the maximum deformation of the body be δ as indicated in Fig.2.1.

Let P_e be the equivalent load, which causes the same stress and deformation in the bar as the impact load P. Assuming that there is no loss of energy at impact, the work done by forces P and P_e should be equal. The work is done by force P through a distance of $(h+\delta)$, while the work done by equivalent force P_e applied gradually will be $(P_e\,\delta/2)$. Thus, we have

$$P\,(h+\delta) = P_e\,\delta/2 \tag{2.8}$$

However, we have from Eqn.1.10

$$\delta = P_e\,l/(AE)$$

Substituting for δ in Eqn.2.8, we obtain

$$P\,[h + P_e l/(AE)] = P_e^2 l/(2AE)$$

Rearranging the above equation, we obtain

$$P_e^2 - 2PP_e - 2P\,hAE/l = 0$$

Solving this equation for P_e, we obtain

$$P_e = P + [P^2 + 2P\,hAE/l]^{0.5}$$

Simplifying the above equation, we have

$$P_e = P\,[1.0 + \{1.0 + 2hAE/(P\,l)\}^{0.5}] \tag{2.9}$$

The negative root is not applicable, since impact increases the effect of the applied load. It can be noted from the above equation that the equivalent load decreases with the deformability of the body (the larger the deformation under the applied load the smaller will be the impact load P_e).

The stress and deformation of the member can be computed from the value of P_e. It should be noted that if the height of fall is zero, the equivalent load (applied gradually) is twice that of the load applied suddenly (Eqn.2.9). Thus, a sudden load causes twice the stresses and deflections induced when the same load is applied gradually on a member. An impact load increases the stresses and deformations many times at the instant of load compared to the same load applied gradually. Equation 2.9 can also be expressed in terms of kinetic energy of the falling body. The deformation of the body under load P applied gradually is denoted by δ_{st}, the static deformation; the static deformation can be expressed as

$$\delta_{st} = P\,l/(AE) \qquad [2.10(a)]$$

Multiplying both sides of Eqn.2.9 by $[l/(AE)]$, and noting that $\delta = [P_e l/(AE)]$, we have

$$\delta = [\delta_{st} + \{\delta_{st}^2 + 2h\delta_{st}\}^{0.5}] \qquad [2.10(b)]$$

The above equation can be re-written as

$$\delta = \delta_{st}[1.0 + \{1.0 + 2h/\delta_{st}\}^{0.5}] \qquad [2.10(c)]$$

If the fall of the impact load is much larger than the static deformation of the body ($h >> \delta_{st}$), Eqn.2.10(b) can be simplified as

$$\delta = [2\,h\,\delta_{st}]^{0.5} \qquad [2.11(a)]$$

The stress in a prismatic body due to the deformation given by the above equation can be expressed as (from Eqn.2.10 (a))

$$\sigma_e = [\delta E/l] = [\{2E/(A\,l)\}(P\,h)]^{0.5} \qquad [2.11(b)]$$

The velocity v, of the impact load, can be expressed as

$$v = [2\,g\,h]^{0.5} \qquad [2.11(c)]$$

Substituting for h from the above equation in Eqn.2.11(b), we have

$$\sigma_e = [\{2E/(A\,l)\}\{(P\,v^2/(2g)\}]^{0.5} \qquad [2.12]$$

The parameter in the second braces of the above equation can be noted to be the kinetic energy of the load at impact [K.E. = $P\,v^2/(2g)$]. It can be noted from Eqn.2.12 that the stress due to impact load is proportional to the kinetic energy of the applied load and the Young's modulus of the material. The larger the Young's modulus the greater will be the stress due to impact load. It can also be noted that the stress is inversely proportional to the volume of the body ($A\,l$); for a given area of cross section, a longer bar develops smaller stress than a shorter bar. This is in contrast to the stress under static (gradually applied) load, where the stress is independent of the length and Young's modulus of the body. However, it is easy to note that the longer body can absorb more energy than a shorter body of the same material and cross section; the capacity to absorb more energy results in smaller stresses.

Referring to Eqn.2.10(b) again, it can be noted that, if $h = 0$ (suddenly applied load), the deformation of the body will be twice that due to static load ($\delta = 2\delta_{st}$).

It was assumed in the above discussions that the body remains within its elastic limit during the application of impact loading. The analysis will be more complicated beyond the elastic limit of the material, because the deformation of the body is no longer proportional to the applied load. The stress can be determined from the stress–strain diagrams (Fig.2.3) by equating the work done by the body to the strain energy of the system [$P(h+\delta)$ = area OABC × volume]. It should be noted that, in this case, the characteristics of the body are assumed to be independent of the speed of straining the material.

The ratio (P_e/P) is some times referred to as the *impact factor.* Impact factor represents the increase in the effect of the load due to its falling on the body (increase in stress and deflection). It can be verified that the expressions in the square brackets of Eqns.2.9 and 2.10(c) are the same; these represent the impact factor of the load for the given height of fall. The significance of work and energy principles, and of impact loading is illustrated with the help of a few examples.

Problem I Determine the modulii of resilience and toughness for mild steel with a proportionality limit of 220.0 MPa and a fracture strain of 26 percent. Assume Young's modulus to be 195.0 GPa.

Solution I By definition, resilience modulus of a material is the area under the stress–strain curve upto the proportionality limit. From Eqn.2.6(b), we have resilience modulus

$$u_r = [\sigma_p^2/(2E)] = [220.0^2/(2 \times 195.0 \times 10^3)] \text{ MNm/m}^3 = 0.124 \text{ MNm/m}^3$$

Toughness modulus is the area under the stress–strain curve upto fracture. The idealised curve shown in Fig.2.3(c) can be adopted for simplicity of computations (the area under the actual curve can be determined using graphical method, if necessary). Referring to Fig.2.3(c), and assuming that proportionality and yield limits are the same for mild steel, we have

$$\begin{aligned}\text{Toughness modulus} &= \text{area OAFG} = \text{area OAC} + \text{area CAFG} \\ &= u_r + (\varepsilon_f - \varepsilon_y)\sigma_y = 0.124 + \{0.26 - 220.0/(195.0 \times 10^3)\} \times 220.0 \\ &= 57.076 \text{ MNm/m}^3\end{aligned}$$

The toughness modulus can also be computed for ductile materials by multiplying ε_f and σ_y without serious error.

Problem II Compare the strain energies of the bars shown in Fig.2.4 when subjected to (i) equal stresses, and (ii) equal loads.

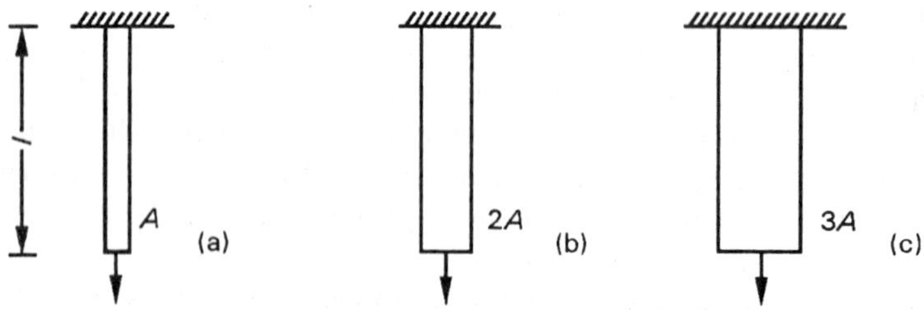

Fig.2.4 Strain energy in bars of different areas (Problem II)

Solution II The bars are of the same material and length but of different areas of cross sections. Strain energy in a body can be computed from Eqn.2.4; the subscripts a, b and c indicate the bars referred.

(i) Bars under equal stress We have (Eqn.2.4(c))

$$U_a = \{\sigma^2/(2E)\}(Al); \quad U_b = \{\sigma^2/(2E)\}(2Al); \quad U_c = \{\sigma^2/(2E)\}(3Al)$$

Thus, we have

$$U_a : U_b : U_c = 1 : 2 : 3$$

The strain energies are in direct proportion to their areas of cross section in this case.

(i) Bars under equal loads Substituting for the appropriate values of areas of cross sections, we obtain (Eqn.2.4(b))

$$U_a = \{P^2l/(2AE)\}; \quad U_b = \{P^2l/(4AE)\}; \quad U_c = \{P^2l/(6AE)\}$$

Thus, we have

$$U_a : U_b : U_c = 1 : 0.50 : 0.33$$

The strain energies are in inverse proportion to their areas of cross section in this case.

Problem III Compare the strain energies in the bars with grooves shown in Fig.2.5 when subjected to maximum permissible stress.

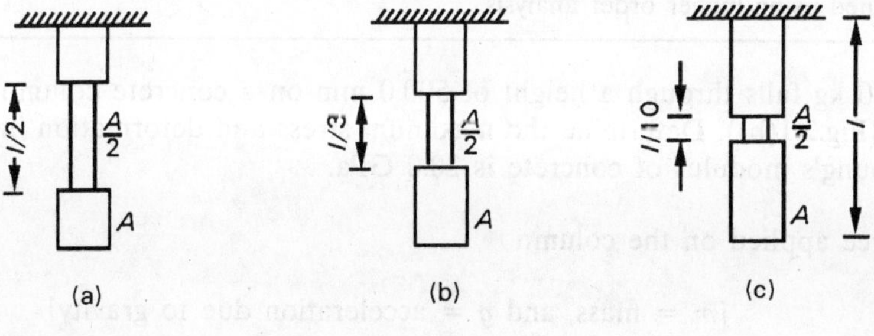

Fig.2.5 Strain energy in bar with grooves

Solution III The bars have grooves of different lengths; the cross sections of the main segments and grooves are the same in all the bars. The bars will be subjected to maximum stress in the segment with reduced cross section; the main segment will be subjected to half the stress in the grooved segment. The strain energy can be computed for each bar considering the lengths of different segments; we have

$$\begin{aligned} U_a &= \{\sigma^2/(2E)\}\{(A/2)(l/2)\} + \{(\sigma/2)^2/(2E)\}\{A(l/2)\} \\ &= (3/8)[\{\sigma^2/(2E)\}(Al)] \end{aligned}$$

$$U_b = \{\sigma^2/(2E)\}\{(A/2)(l/3)\} + \{(\sigma/2)^2/(2E)\}\{A(2l/3)\}$$

$$= (1/3)[\{\sigma^2/(2E)\}(Al)]$$

$$U_c = \{\sigma^2/(2E)\}\{(A/2)(l/10)\} + \{(\sigma/2)^2/(2E)\}\{A(9l/10)\}$$

$$= (11/40)[\{\sigma^2/(2E)\}(Al)]$$

Thus, we have

$$U_a \ : \ U_b \ : \ U_c = 1 \ : \ 0.89 \ : \ 0.73$$

The parameter within the square brackets above indicates the strain energy in a uniform bar of area A and length l subjected to uniform stress σ. It can be noted that the strain energy of a bar is reduced significantly as the length of the groove is reduced; the smaller the length of the groove the smaller will be the strain energy for a given maximum stress in the bar. If the area of the bar is reduced by half uniformly over its entire length, the strain energy of the bar will be reduced by half. On the other hand, if the area of the bar is reduced by half over half its length, the strain energy is reduced to 0.375 times that of the original bar, and to 0.275 times when the length of the groove is one-tenth the bar length.

Note Since strain energy denotes the impact resistance of the body, it can be concluded that the smaller the length of groove the smaller will be the impact resistance of the bar. That is, even a small reduction in area over a small length of the bar will cause failure at a much smaller impact load compared to the uniform bar. In addition, a sudden reduction of area of a bar results in a non-uniform stress higher than that given by the elementary equation (P/A); this phenomenon is known as *stress concentration* and can be determined using higher order analysis.

Problem IV A mass of 200.0 kg falls through a height of 500.0 mm on a concrete column of 300.0 × 400.0 mm section (Fig.2.1(b)). Determine the maximum stress and deformation in the 4.0 m long column, if Young's modulus of concrete is 20.0 GPa.

Solution IV The impact force applied on the column

$$P = m\,g \qquad [m = \text{mass, and } g = \text{acceleration due to gravity}]$$

$$= 200.0 \times 9.81 \text{ N} = 1\ 962.0 \text{ N}$$

From Eqn.2.9, we have the equivalent force on the column given by

$$P_e = P\,[1.0 + \{1.0 + 2hAE/(P\,l)\}^{0.5}]$$

$$= 1\ 962.0[1.0 + \{1.0 + 2 \times 500.0 \times (300.0 \times 400.0) \times 20.0 \times 10^3/(1\ 962.0 \times 4\ 000.0)\}^{0.5}]$$

$$= 1\ 962.0[554.00] \text{ N} = 1\ 086.95 \text{ kN}$$

$\sigma_e = P_e/A$ [equivalent stress]

$= -1\ 086.95 \times 10^3/(300.0 \times 400.0) = -9.06$ Nmm^{-2}

Elongation of the bar is obtained by

$$\delta = \sigma_e\, l/E = -9.06 \times 4\ 000.0/(20.0 \times 10^3) = -1.81 \text{ mm}$$

Negative sign indicates compressive stress and shortening. The compressive stress on the column section, when the load is applied gradually, works out to be 0.016 Nmm^{-2}, and the corresponding shortening will be 0.003 3 mm. That is, the stress and deformation under the impact load are 554.0 times the corresponding values, when the load is applied gradually.

Problem V A mass of 50.0 kg falls through a height of 60.0 mm on the collar of a suspended steel bar of 2.0 m length (Fig.2.1(a)). Determine the stress and deformation, when the bar diameter is
(i) 20.0 mm over the entire length and (ii) 20.0 mm over half the length and 12.0 mm over the other half.

Solution V Since it is a steel bar, $E = 200.0$ GPa. Impact load $P = 50.0 \times 9.81 = 490.5$ N.

(i) Bar of uniform section Area of cross section $A = 314.16$ mm^2. The equivalent force on the bar is given by Eqn.2.9.

$$P_e = P\,[1.0 + \{1.0 + 2hAE/(P\,l)\}^{0.5}]$$

$$= 490.5[1.0 + \{1.0 + 2 \times 60.0 \times (314.16) \times 200.0 \times 10^3/(490.5 \times 2\ 000.0)\}^{0.5}]$$

$$= 490.5\,[88.675] \text{ N} = 43.495 \text{ kN}$$

Stress in the bar at the instant of load application is given by

$$\sigma_e = P_e/A = 43.495 \times 10^3/(314.16) = 138.45 \text{ Nmm}^{-2}$$

Extension of the bar can be computed as

$$\delta = \sigma_e\, l/E = 138.45 \times 2\ 000.0/(200.0 \times 10^3) = 1.38 \text{ mm}$$

(ii) Bar of varying cross section Let the cross-sectional areas be A_1 and A_2. The extension of the bar, when an equivalent force P_e is applied gradually, is the sum of the extensions of the two segments. The extension of the bar under P_e can be expressed as ($A_1 = 314.16$ mm^2, $A_2 = 113.10$ mm^2 and $l_1 = l_2 = 1\ 000.0$ mm)

$$\delta \;=\; P_e\, l_1/(A_1 E) + P_e\, l_2/(A_2\, E) = P_e[(l_1/A_1) + (l_2/A_2)]/E$$

$$=\; P_e[(1\ 000.0/314.16) + (1\ 000.0/113.10)]/200.0 \times 10^3 = [60.124 \times 10^{-6}]P_e$$

Since the bar is of varying section, Eqn.2.9 cannot be applied directly; the energy equation (Eqn.2.8) should be applied in this case.

$$P(h+\delta) = P_e\,\delta/2 \qquad \text{(Eqn.2.8)}$$

$$490.5(60.0 + 60.124 \times 10^{-6}P_e] = P_e^{\,2}(60.124 \times 10^{-6})/2$$

Solving the above equation, we obtain

$$P_e = 31.783 \text{ kN}$$

Extension of the bar $\delta = [60.124 \times 10^{-6}]P_e = 1.91$ mm

Stress in the larger section $= P_e/A_1 = 101.17$ Nmm^{-2}

Stress in the smaller section $= P_e/A_2 = 281.02$ Nmm^{-2}

Note the changes in the stresses and in the elongation of the bar due to reduced section over half its length.

Problem VI A lift is operated by two ropes of 30.0 m length comprising 36 wires of 1.2 mm diameter each. The cage weighs 1.0 kN and the self-weight of the rope is 3.2 Nm^{-1}. Determine the maximum load the lift can carry if it
(i) operates without any drop, and (ii) drops 100.0 mm during operations.
Assume Young's modulus for rope as 75.0 GPa and an allowable stress of 120.0 MPa.

Solution VI Area of cross section of rope wires

$$A = 2 \times 36 \times \pi(1.2^2)/4 \text{ mm}^2 = 81.43 \text{ mm}^2$$

Maximum stress in the rope occurs at its top end due to the weight of the cage and self-weight. This stress is in addition to the stress caused by the load carried by the lift. The initial stress σ_1 can be computed as

$$\sigma_1 = (1.0 \times 10^3 + 2 \times 30.0 \times 3.2)/81.43 \text{ Nmm}^{-2} = 14.64 \text{ Nmm}^{-2}$$

The load carried by the lift can be termed as live load. The maximum equivalent live load the lift can carry is limited by the maximum stress in the ropes, namely 120.0 MPa. Maximum equivalent live load on the ropes can be computed as

$$P_e = (120.0 - 14.64) \times 81.43 \text{ N} = 8.58 \text{ kN}$$

Maximum extension of the rope can be computed as

$$\delta_{max} = [(120.0 - 14.64) \times 30.0 \times 10^3/(75.0 \times 10^3)] \text{ mm}$$

$$= 42.14 \text{ mm}$$

(i) Operation without any drop The application of load on the lift is assumed to be sudden. Let P_e be the equivalent live load on the lift. The work done by the equivalent live load and the actual live load should be equated (Eqn.2.8); when $h = 0$, this yields

$$P\,\delta = P_e\,\delta/2 \quad \text{or} \quad P = P_e/2$$

$$P = [8.58/2]\ \text{kN} = 4.29\ \text{kN}$$

(ii) Operation with a 100 mm drop The work equation (Eqn.2.8) can be written as

$$P\,(h + \delta_{\max}) = P_e\,\delta_{\max}/2$$

Substituting the given data and the values computed so far in the above equation, we obtain

$$P\,(100.0 + 42.14) = 8.58 \times 42.14/2 \qquad [P_e = 8.58\ \text{kN}]$$

or $\quad P = 1.27$ kN

Thus, the lift can carry a live load of 4.29 kN without any drop and 1.27 kN with a drop of 100 mm.

Problem VII A load P is lowered by a steel rope of cross-sectional area A at a constant velocity v. If the upper end of the rope is stopped suddenly when the length of the rope is l, compute the maximum stress and deformation in the rope.

Solution VII The problem can be analysed by applying the principle of conservation of energy. The total energy of the system remains the same through the operations. The energy of the system at the instant the rope comes to a halt is equated to the strain energy of the system at the instance the rope is stopped. The strain energy of the sytem at the instant the rope comes to a halt can be expressed as

$$U = P_e\,\delta/2 \qquad \text{[VII.1]}$$

where P_e = the maximum force developed in the rope and δ = the maximum deflection of the rope.

The load was moving with a velocity v at the instance the rope was stopped at its upper end. Let δ_{st} be the static deflection of the rope under the load $P\,[\delta_{st} = P\,l/(AE)]$. Thus, the load was at a higher level when the rope was stopped than when the load came to a halt by an amount $(\delta - \delta_{st})$. The energy of the system when the rope was stopped comprised the strain energy of the rope, the kinetic energy of the load and its potential energy; the energy can be expressed as

$$W = (P\,\delta_{st}/2) + \{(P\,v^2)/(2g)\} + P\,(\delta - \delta_{st}) \qquad \text{[VII.2]}$$

The weight of the rope is neglected in this formulation. Equating the energies given by Eqns.VII.1 and VII.2, we obtain

$$(P_e\,\delta/2) = (P\,\delta_{st}/2) + [(P\,v^2)/(2g)] + P\,(\delta - \delta_{st})$$

Substituting for δ and δ_{st} in the above equation, we obtain

$$[P_e^2 l/(2AE)] = [P^2 l/(2AE)] + [(P v^2)/(2g)] + P[\{P_e l/(AE)\} - \{P l/(AE)\}]$$

Multiplying both sides of the equation by $(2AE/l)$, and simplifying, we obtain

$$P_e^2 - 2PP_e + P^2 = P v^2 AE/(g\, l)$$

The above equation can be simplified further as

$$P_e - P = [P v^2 AE/(g\, l)]^{0.5}$$

or $$P_e = P + [P v^2 AE/(g\, l)]^{0.5} \qquad \text{[VII.3]}$$

The maximum stress and deflection in the rope can be computed using the above equation.

2.4 Summary

It is usual to assume gradual application of loads on a structure unless specified otherwise. Sudden application of loads induces higher stresses in a structure than when the same load is applied gradually; the stresses increase several times when the same load is dropped on a body (impact loading). Several structures, especially machines and bridges, are subjected to impact loads. It is thus essential to comprehend the effects of sudden and impact loads. The effects of impact loads (which include sudden loading) can be computed from the principle of conservation of energy. The concepts of work (external and internal) are explained in this chapter. The stresses and deformation of structures subjected to impact loads are compared with those due to the same load applied gradually; it is shown that the stresses due to impact loads are several times those due to the latter.

The concepts of strain energy, resilience modulus and toughness modulus are also explained.

Strain energy $= (1/2) \int [\varepsilon\, \sigma]\, dv$

$= (1/2)[\varepsilon\, \sigma] V$ (when ε is constant)

$= [\sigma^2/(2E)] V$

Resilience modulus = area of the stress–strain curve from the origin upto proportionality limit

$= (1/2)[\sigma_p\, \varepsilon_p]$

Toughness modulus = area of the stress–strain curve from the origin upto fracture

$= \sigma_p\, \varepsilon_f$

Equivalent load P_e $= P\left[1.0 + \{1.0 + 2hAE/(P\, l)\}^{0.5}\right]$

2.5 Exercises

2.1 Explain the concepts of work done by a load applied on a body (i) gradually, (ii) suddenly, and (iii) with impact.

2.2 What is strain energy of a material? What is its significance?

2.3 Discuss the influence of a groove on the performance of a bar in tension.

2.4 Discuss the significance of resilience modulus and toughness modulus in evaluating the suitability of materials in structural design.

2.5 What is impact loading? Why does an impact load induce higher stress in a body than when the same load is applied gradually? Explain the concept of impact factor.

2.6 Why does a body with lower value of Young's modulus develop lower stress under impact load than a similar body with higher value of Young's modulus?

2.7 A prismatic steel bar of length l and cross-sectional area A is suspended freely from its upper end. Calculate the strain energy of the bar, if the unit weight of steel is w.
[*Hint:* Integrate the expression for the strain energy of a segmental length dx at a distance x from the free end over its entire length.]

2.8 Compare the modulii of resilience and toughness for mild steel, brass, aluminium and cast iron from the data given in Table 1.1 (Chapter 1) using idealised bilinear stress–strain curves.

2.9 Compute the strain energy in a steel bar (E = 200.0 GPa) of length 2.7 m and 22.0 mm diameter under a load of 50.0 kN. What is the resilience modulus of the bar, if the yield stress is 240.0 MPa?

2.10 Determine the resilience and toughness modulii of mild steel (E = 200.0 GPa) with a yield stress of 250.0 MPa and fracture strain of 28.5 percent. Neglect strain hardening effects. From these data determine the impact resistance of a bar of 12.0 mm diameter and 500.0 mm length.

2.11 Three bars of lengths 2.0 m, 2.9 m and 3.6 m are to be used as tension members subjected to possible impact loading. Investigate the suitability of the bars when their diameters are 30.0 mm, 25.0 mm and 20.0 mm, respectively. What would be the choice, if the first two bars have 1.1 and 1.0 mm deep grooves, respectively over a length of 10.0 mm?

2.12 Compare the impact resistance of a steel bar of diameter 12.0 mm and 1.5 m length with that of an aluminium bar of diameter 20.0 mm and 2.5 m length. Assume the Young's modulus of steel as 190.0 GPa and of aluminium as 85.0 GPa. The yield stress of steel is 240.0 MPa with an ultimate strain of 25.0% and the corresponding values for aluminium are 120.0 MPa and 22.0%.

2.13 A mass of 250.0 kg falls through a height of 300.0 mm on a concrete column of 230.0 × 500.0 mm section. Determine the maximum stress and deformation in the 4.5 m long column, if Young's modulus of concrete is 20.0 GPa.

2.14 A mass of 60.0 kg falls through a height of 50.0 mm on the collar of a suspended steel bar of 3.0 m length. Assuming E = 200.0 GPa for steel, determine the maximum stress, deformation and impact factor, when the bar diameter is
(i) 30.0 mm over the entire length and (ii) 30.0 mm over half the length and 15.0 mm over the other half.

2.15 Determine the height through which the mass of 150.0 kg may fall on the collar of a 2.7 m long suspended steel bar of diamater 25.0 mm so that the instantaneous stress does not exceed 415.0 MPa. Assume E = 200.0 GPa for steel.

2.16 What is the maximum height of fall for a mass of 50.0 kg on a 3.6 m long mild steel bar of 20.0 mm diameter so that the instantaneous stress does not exceed 250.0 MPa. Assume E = 190.0 GPa.

2.17 Design a 1.5 m long steel bar to sustain a load of 5.0 kN falling on it through a height of 100.0 mm for a maximum allowable stress of 140.0 MPa;
(i) neglect the extension of the bar and (ii) consider the bar extension as well.
Assume E = 210.0 GPa. [*Hint*: Determine the area of the bar required from Eqn.2.9 for the given constraints, and choose a suitable diameter for the bar.]

2.18 A lift is operated by two ropes of 40.0 m length comprising 36 wires of 1.6 mm diameter each. The cage weighs 1.5 kN and the self-weight of rope is 5.7 Nm^{-1}. Determine the maximum load the lift can carry, if it
(i) operates without any drop, and (ii) drops 150.0 mm during operations.
Assume Young's modulus for rope to be 150.0 GPa and an allowable stress of 500.0 MPa.

2.19 A mass of 500.0 N falls on the collar of a 20.0 mm diameter steel bar from a height of 100.0 mm. Investigate whether the bar can withstand the load. Assume a yield stress of 275.0 MPa, a fracture strain of 24.5 percent and $E = 195.0$ GPa for steel.

2.20 A lift of mass 2 000.0 kg comes to a sudden halt while moving with a downward velocity of 1.2 ms^{-1}. The lift is supported by two ropes, each comprising 12 steel wires of 3 mm diameter. Compute the maximum stress and deflection in the ropes, if the length of the ropes is 25.0 m when the lift stopped. Assume the Young's modulus of the ropes to be 120.0 GPa.

2.21 A weight P attached to a string of length l, has cross section A and Young's modulus E. One end of the string is fixed to a support and the weight is allowed to fall freely over the full length of the string. Show that the maximum force on the string is $P[1 + (1 + 2EA/P)^{0.5}]$ when the weight of the string is neglected.

2.22 A 1.8 m long compound bar comprising a steel rod of 25.0 mm diameter rigidly encased in a brass tube of external diameter 30.0 mm and 25.0 mm internal diameter is subjected to a load of 200.0 kg falling through 60.0 mm. Determine the stresses in the rod and tube, and the deflection [*Hint*: Deformations of brass and steel elements are equal. Apply the energy principles to determine the equivalent load.]

3 Shear stresses

3.0 Introduction

While a direct force is applied normal to the surface of a body, a shear force (S.F.) is applied on the surface itself as indicated in Fig.3.1(a). Shear force is often encountered in structures, and tends to cause sliding of a part of the body over the other. Whereas direct forces cause deformations along the axis of application, S.F. tend to distort the member. The deformation caused by S.F. is indicated in Fig.3.1(b); the angles of the body are altered due to S.F.

Shear forces occur in truss joints (riveted, cottered and welded) connecting various members, in beams subjected to bending, and in shafts subjected to torsion. Shear transferred directly from one member to another is known as *direct shear*, and occurs at truss joints (it is referred to as *pure shear*, sometimes). The body of Fig.I.5(d) is subjected to S.F. at section X–X as shown in the F.B. diagram; such a S.F. is known as *flexural shear* (flexure means bending). Shear is also developed in members due to torsion (Fig.I.5(e)), and is known as *torsional shear*.

Some of the aspects of direct S.F. and its effects on a member are discussed in this chapter. Other types of shear are presented in subsequent chapters. Stresses developed in joints of steel structures are discussed briefly. Various types of joints and their design are explained. The concepts of riveted, bolted and welded joints are also discussed. Shear stresses often occur in combination with direct stresses; compound stresses are discussed in chapter 7.

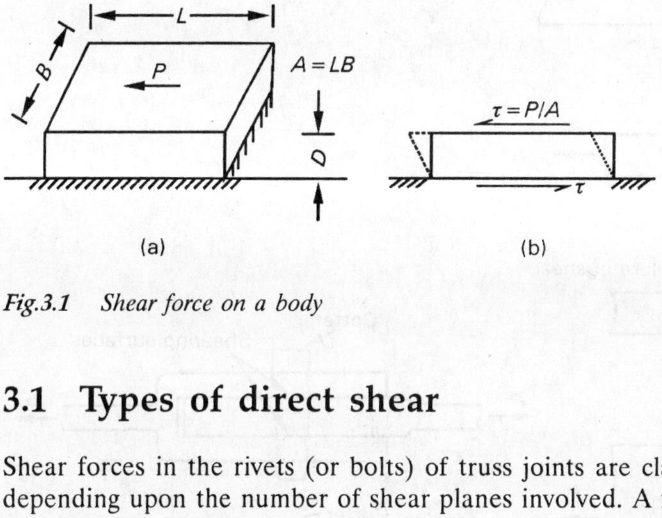

Fig.3.1 Shear force on a body

3.1 Types of direct shear

Shear forces in the rivets (or bolts) of truss joints are classified into single and double shear depending upon the number of shear planes involved. A few examples of S.F. in riveted joints are indicated in Fig.3.2.

The plates of the joint shown in Fig.3.2(a) are lapped and connected by means of rivets passing through matching holes in the plates. Such a joint is known as a *lap joint*. In the absence of rivets, the plates slide off at the interface and cannot sustain any force. Only one rivet is indicated in the figure but several rivets may be provided in practice. The force at the joints is transferred from one plate to the other through the force in the rivets at the interface. This is sometimes known as *direct shear*. The force is assumed to be shared equally by all the rivets and the frictional resistance between the sliding surfaces is ignored. The possible failure mode is also indicated in the figure. Only one shear plane exists in a lap joint (the interface of the plates). In such a case the rivets are subjected to shear on only one plane, and are in *single shear*. The shearing (failure) plane is circular, of diameter *d* (the diameter of the rivet).

The plates of the joint indicated in Fig.3.2(b) butting against each other are provided with cover plates and connected by means of rivets; such joints are known as *butt joints*. The figure also indicates possible failure mode of the rivets. The rivets in a butt joint are subjected to shearing forces along two planes as indicated in Fig.3.2(b); such rivets are said to be in *double shear*. The segments of the rivets above and below the interfaces of the sliding plates tend to shear off. The shearing surfaces are circular of diameter *d* in such cases, *d* being the diameter of the rivets.

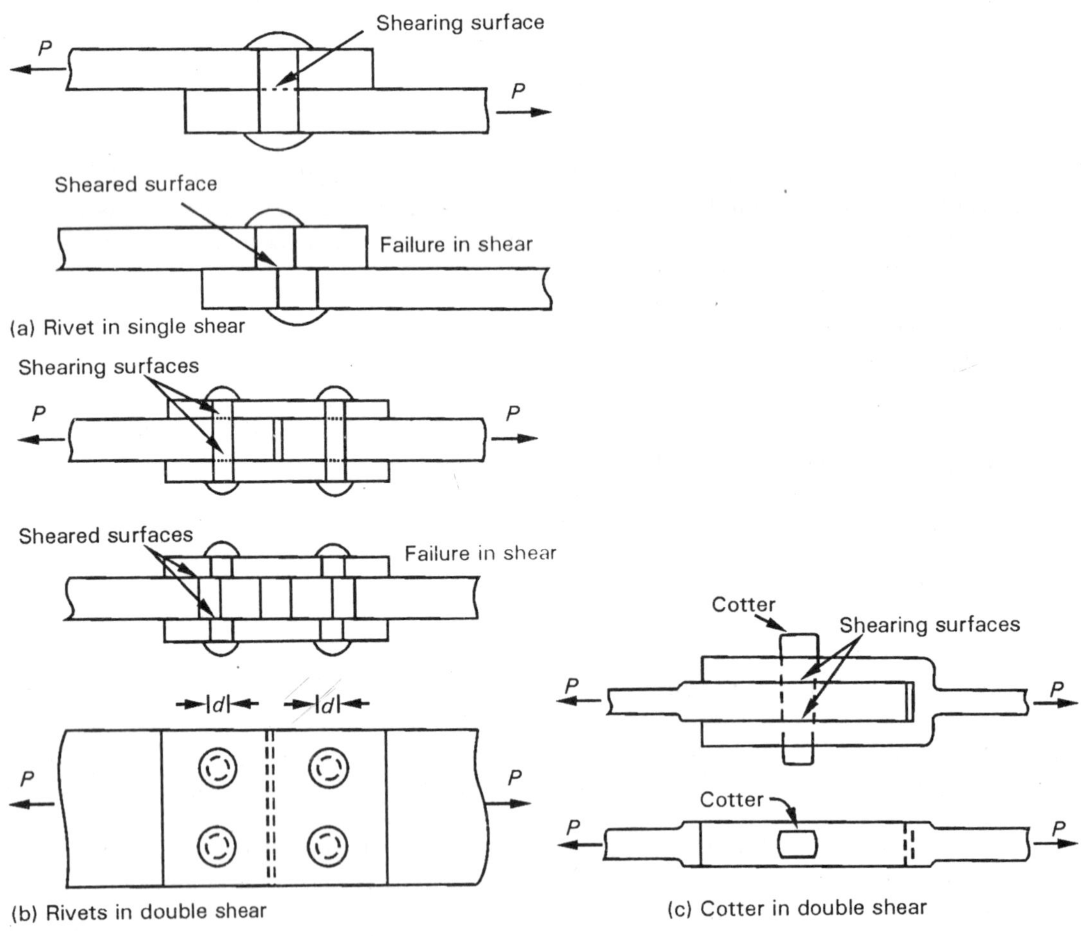

Fig.3.2 Typical joints

The cotter pin of Fig.3.2(c) is also in double shear; the shearing surfaces may be rectangular or circular in a cottered joint depending upon the cross section of the cotter. Fig.3.2(c) shows a tapering cotter of rectangular section. Fig.3.3 indicates shear in a plate due to punching; the punch of diameter d is forced through the plate to make a hole. Such a S.F. is known as *punching shear*. The shearing surface in this case is a cylinder of diameter d and length t, the thickness of the plate.

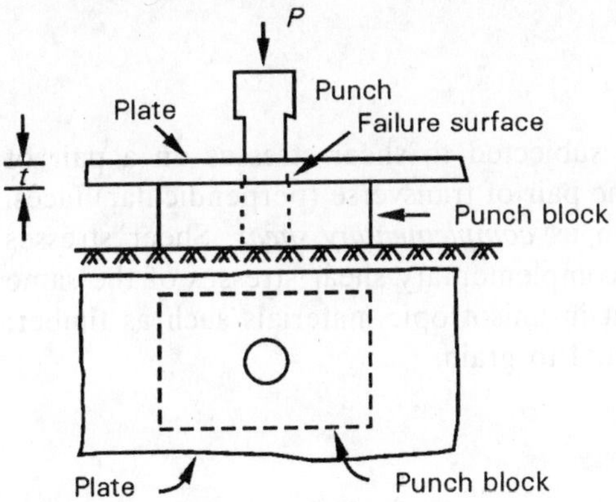

Fig.3.3 Punching of holes in a steel plate

3.2 Shear stress

Shear force is assumed to be distributed uniformly on the surface on which it is applied. Shear stress is defined as the force per unit area, the same as direct stress, and is denoted by the Greek letter τ (tau). Referring to Fig.3.1, shear stress can be expressed mathematically as

$$\tau = P/A \qquad [3.1(a)]$$

$$= dP/dA \qquad [3.1(b)]$$

where A is the area over which the S.F. acts.

The first of the above equations refers to average stress on a surface, while the second refers to shear stress at a point in the case of varying shear stress. Shear stress has the same dimensions as direct stress, namely force per unit area. In SI units, shear stress is usually expressed in Nmm^{-2} or MPa.

3.3 Complementary shear

It should be noted that S.F. comprises a pair of equal and opposite parallel forces applied on the opposite surfaces of a body as indicated in Fig.3.1. Considering the prismatic body shown in Fig.3.1(a), force P applied on the upper surface of the body induces a shear stress $\tau = [P/(L \times B)]$, where L and B are the dimensions of the surface sustaining the S.F.

An equal and opposite force should act on the lower surface of the body in order to maintain equilibrium of the horizontal forces. However, the forces due to stresses form an anti-clockwise couple of magnitude PD (where D is the thickness of the body). The body cannot thus be in equilibrium unless a clockwise couple of magnitude PD also acts on the body. The clockwise couple should comprise forces acting on the vertical faces of the body inducing a shear stress of magnitude τ' as indicated in Fig.3.4. The condition of moment equilibrium can be written as

$$(\tau LB)D = (\tau' DB)L \quad \text{or} \quad \tau = \tau'$$

The above equation shows that when a body is subjected to shear stresses on a pair of opposite faces, the same shear stress is induced on the pair of transverse (perpendicular) faces. The shear on the transverse pair of faces is known as *complementary shear.* Shear stresses cannot exist alone, and are always accompanied by complementary shear stresses of the same magnitude. Complementary shear may be significant in anisotropic materials such as timber; timber is weaker in shear along the grain than normal to grain.

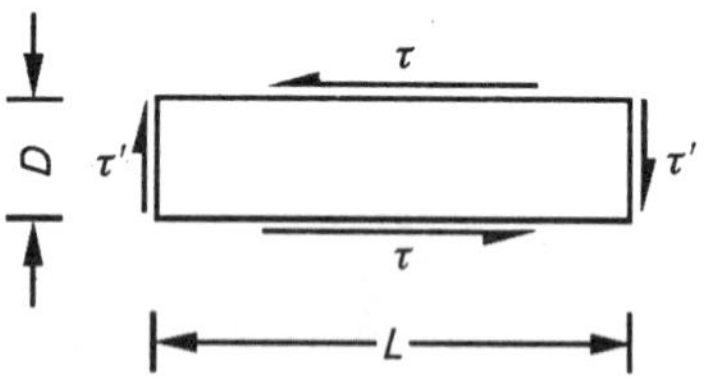

Fig.3.4 Complementary shear stress

3.4 Shear strain

Shear stresses cause distortion of a body as indicated in Fig.3.1(b). The upper face of the body slides relative to the lower surface; the bottom surface is assumed to be fixed in the figure. Such a distortion causes the corner angles to be altered by γ (Greek letter, gamma) as indicated in Fig.3.5. *Shear strain* is defined as the change in a right angle of a body, measured in radians, due to applied shear stress. Shear strain is non-dimensional, being the angular change in a body.

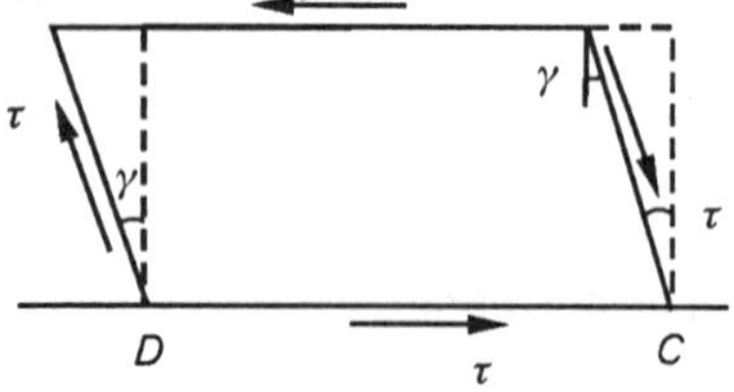

Fig.3.5 Shear stress and strain

3.5 Rigidity modulus

Shear strain in an elastic body is proportional to the applied shear stress within the elastic limit of the material. The constant of the proportionality of shear stress and strain is known as the *rigidity modulus* of the material (it is also known as *shear modulus*). Rigidity modulus is denoted by G and has the same units as Young's modulus (force per unit area). Mathematically, we have

$$G = \tau/\gamma \qquad [3.2]$$

Rigidity modulus is a function of Young's modulus and Poisson's ratio of a material. This relation will be derived in the next chapter on stress analysis.

Problem I Determine the shear stress in the rivets of a lap joint subjected to a force of 100.0 kN. The plates are connected by four rivets of 20.0 mm diameter.

Solution I The rivets are subjected to a force of 100.0 kN. The force is assumed to be equally distributed among the four rivets of the joint ($d = 20.0$ mm). The rivets in a lap joint are in single shear (Fig.3.2(a)).

Force per rivet $P = 100.0/4 = 25.0$ kN

Area of cross section of a rivet $A = \pi d^2/4 = 314.16\ \text{mm}^2$

Shear stress $= P/A = 25.0 \times 10^3/314.16 = 79.6\ \text{Nmm}^{-2}$

Problem II Design the rivets of a butt joint to sustain 500.0 kN. Assume an allowable stress τ_s of 90.0 MPa.

Solution II The rivets in a butt joint are in double shear (Fig.3.2(b)).

Area of rivets required $A = P/\tau_s = 500.0 \times 10^3/90.0 = 5\ 555.6\ \text{mm}^2$

If a is the area of cross section of a rivet, the number of rivets required N (the rivets are in double shear) is given by

$$N = A/(2a)$$

The above equation has two unknown parameters (N and a); the number of rivets can be determined by assuming a suitable diameter for rivets. Assuming 22.0 mm diameter rivets, we have

$$a = \pi(22.0^2)/4 = 380.1\ \text{mm}^2$$

Thus, we obtain

$$N = 5\ 555.6/(2 \times 380.1) = 7.3 \qquad \text{(or, say, 8)}$$

Provide eight rivets of 22.0 mm diameter on each side of the joint.

3.6 Strain energy

Strain energy in a body subjected to shear stresses can be computed the same way as that for direct stresses discussed in Chapter 2 (Section 2.2). The strain energy in a system within the proportionality limit can be computed by applying Eqns.2.3 and 2.4. We have

Strain energy $U = (1/2)\times$ (couple formed by the S.F.) $\times$ (angle of deformation)

Referring to Figs.3.1 and 3.5, the couple formed by shear stresses is given by $(\tau \times L \times B) \times D$, while the angle of deformation is γ. Substituting in the above equation and noting that $\gamma = \tau/G$, we obtain

$$U = (1/2) \times [(\tau \times L \times B) \times D] \times \gamma$$

$$= (1/2)[\tau\, \gamma]V \qquad \text{[where } V = LBD\text{, the volume of the element.]}$$

$$= [\tau^2/(2G)]V \qquad [3.3(a)]$$

or $$u = [\tau^2/(2G)] \qquad [3.3(b)]$$

3.7 Summary

Shear stresses occur in several structural elements. While direct stresses cause deformations along their directions (elongation or shortening), shear stresses cause distortion of the bodies. Shear stresses are termed direct (or pure), if they are induced by direct S.F.; flexural shear stresses occur due to bending and torsional shear stresses occur due to torsion.

Shear stress is defined as the force per unit area; the ratio of shear stress to shear strain is known as the rigidity (or shear) modulus of the material. Shear strain is defined as the change in a right angle due to the applied shear stress.

Shear stress τ = Shear force/area = P/A [$= dP/dA$, in the differential form]

Rigidity modulus $G = \tau/\gamma$ [γ = shear strain]

The shear in rivets and bolts is generally known as single shear, if shear occurs over only one surface, and as double shear if shear occurs over two surfaces. Some of the aspects of direct S.F. and its effects on a member are discussed in this chapter. Stresses developed in various types of joints (riveted, bolted and welded) in steel structures, and the design of joints are discussed briefly. The concept of shear strain energy is also explained.

Strain energy per unit volume of a body $u = [\tau^2/(2G)]$.

3.8 Exercises

3.1 Explain the difference between shear stresses and direct stresses.
3.2 What is complementary shear? What is its significance in analysis?

3.3 What are single shear and double shear in structural elements? Discuss various types of structural elements subjected to shear.

3.4 Describe various types of joints in structural elements and the stresses induced in them.

3.5 Discuss the concept of shear strain energy and its significance in the material behaviour.

3.6 Determine the force required to punch 65.0 mm diameter holes through a 2.0 mm thick steel plate assuming an ultimate shear stress of 320.0 MPa.

3.7 A steel punch can be operated at a stress of 1 200.0 MPa. Determine the smallest diameter of hole it can punch through a steel plate of 10.0 mm thickness, if the ultimate shear stress is 300.0 MPa. What should be the operating stress, for a hole of 18.0 mm diameter in a 20.0 mm thick plate ?

3.8 A steel punch operates at a force of 275.0 kN. Determine the largest hole it can punch through a 12.0 mm thick plate of 360.0 MPa ultimate shear stress.

3.9 Compute the shear stress in the rivets of a lap joint subjected to a force of 150.0 kN. The plates are connected by six rivets of 16.0 mm diameter.

3.10 Determine the safety factor for the bolt shown in Fig.E 3.1, if the allowable stress in shear is 100.0 MPa. The plates are to sustain a force of 27.0 kN.

3.11 Determine the stress in the bolt shown in Fig.3.1 due to an axial load of 45.0 kN in the plates.

3.12 Three concrete blocks are bonded by epoxy resin over the contact faces as shown in Fig.E 3.2. Determine the interfacial stress, if the applied load is 1.5 MN. What is the load the system can sustain, if the strength of the epoxy joint is 7.5 MPa?

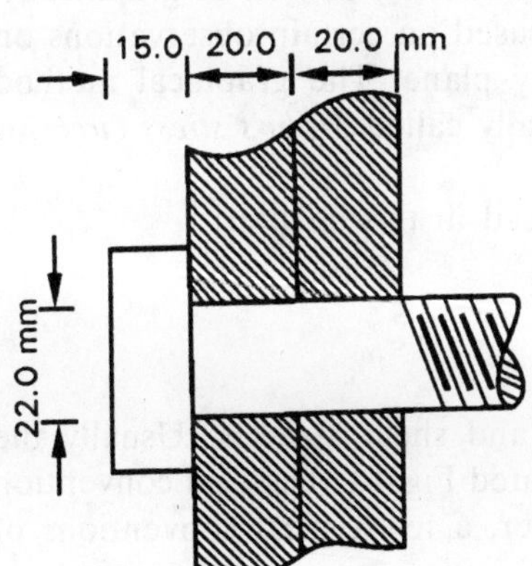

Fig. E 3.1

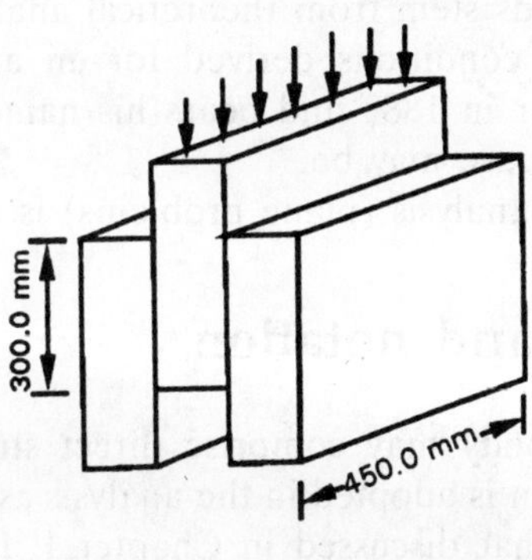

Fig. E 3.2

4 Stress analysis

4.0 Introduction

In the previous chapters, direct and shear forces were considered to act independently. The resultant stresses were determined on planes normal to the direct forces, and on the same plane for S.F. applied on a body. However, direct and shear forces usually act together on a body, and the maximum value of the resultant stress may not occur on the plane of load application. The direction of stresses at various locations in a body may vary depending upon the body configuration and applied loads. It is thus necessary to compute stresses on planes other than those of load application. The process of stress computation on an arbitrary plane, and the determination of the maximum values of stresses and the planes of their occurrence for a given load system is usually termed as *stress analysis*. Similarly, *strain analysis* involves the determination of strains in any direction in a body, and of their maximum values for the given load conditions.

An element of infinitesimal dimensions is assumed in the analysis, so that the stresses approach the conditions at a point in the limit as the dimensions tend to zero. The stresses can be assumed to be distributed uniformly over the surfaces of the element in such a case. The thickness of the element is assumed to be unity in two-dimensional analysis, and the effect of thickness is ignored. Stress and strain analyses can be done analytically as well as graphically. However, graphical methods stem from theoretical analysis, based on certain observations on the expressions for stress conditions derived for an arbitrary plane. The graphical method was developed by O. Mohr in 1882 and bears his name, usually called *Mohr's stress circle* or *Mohr's strain circle* as the case may be.

Two dimensional stress analysis (plane problems) is discussed in this chapter.

4.1 Conventions and notation

Stresses at a point in a body may comprise direct stresses and shear stresses. Usually the Cartesian coordinate system is adopted in the analysis as indicated Fig.4.1. The sign convention is generally the same as that discussed in Chapter I. However, a few typical conventions of stress analysis are explained here.

4.1.1 Direct stresses

Direct (or normal) stresses are designated by the Greek letter σ (sigma) with a subscript in lower case Roman letter to indicate the direction of action. In Fig.4.1(a), σ_x and σ_y indicate the stresses in the directions of the X- and Y-axes, respectively. As in the general convention, tensile stresses are positive and compressive stresses are termed negative.

4.1.2 Shear stresses

The pair of shear stresses on parallel planes that form a clockwise couple are termed positive, and the pair that forms a counter-clockwise couple are termed negative. According to this convention, the shear stresses on vertical faces of the body shown in Fig.4.1 are positive and those on horizontal faces are negative. Shear stresses are denoted by the subscripts of the plane in which they lie (for instance, τ_{xy} denotes shear stress in the X–Y plane and τ_{yz} denotes that in the Y–Z plane).

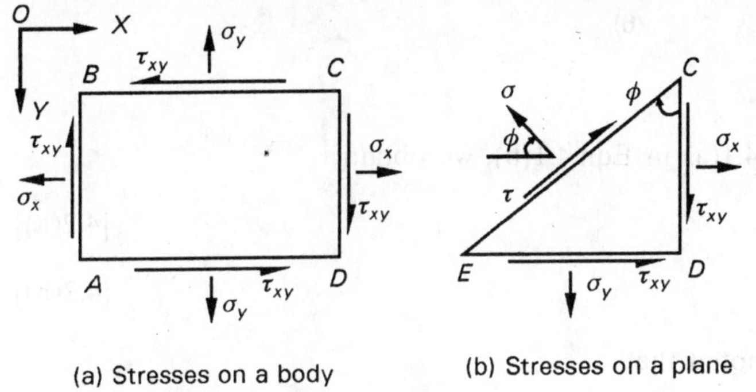

Fig.4.1 Conventions and notation

4.1.3 Inclination of planes

The inclination of planes measured from the Y-axis in clockwise direction is positive; the inclination measured in counter-clockwise direction is negative. The positive direction of angular measurements is indicated in Fig.4.1(b).

4.2 Simple stress condition

The body shown in Fig.4.2(a) is subjected to a simple tensile force P_x. The force on an arbitrary plane EF inclined at ϕ with the Y-axis can be determined from the F.B. diagram shown in Fig.4.2(b). The force P_x on plane EF, required to maintain equilibrium of segment $EFCD$, can be resolved into a normal component P_n and tangential component P_s; the normal component induces direct stress while the tangential component causes shear stress on plane EF. From simple resolution of forces, we have

$$P_n = P_x \cos\phi \quad \text{and} \quad P_s = P_x \sin\phi \qquad [4.1(a)]$$

However, stresses are more significant than forces in the design of structures in order to restrict them to the specified values for a given material. The stresses σ and τ on planes EF and CD can be obtained by dividing the forces by the corresponding areas on which they act. Assuming unit thickness of the body, and a depth of l, we have

$$\sigma_x = P_x/l; \quad \sigma = P_n/(l/\cos\phi) \qquad [EF = (CD/\cos\phi) = (l/\cos\phi)]$$

$$\text{and } \tau = -P_s/(l/\cos\phi) \qquad [4.1(b)]$$

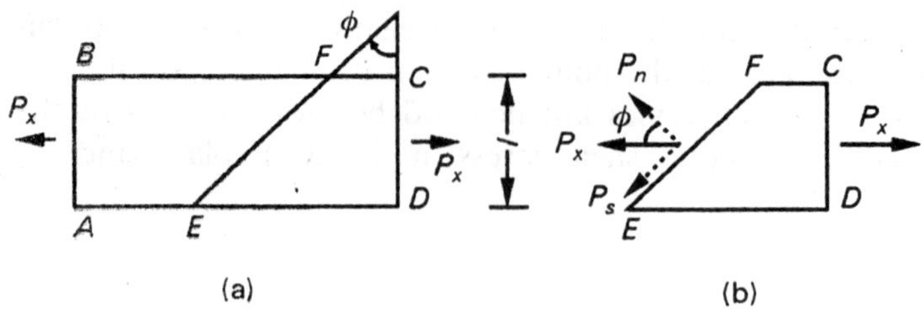

Fig.4.2 An element under simple tension

Substituting for P_n and P_s from Eqn.4.1(a) in Eqn.4.1(b), we obtain

$$\sigma = \sigma_x \cos^2 \phi \qquad [4.2(a)]$$

$$\tau = -\sigma_x \cos\phi \sin\phi \qquad [4.2(b)]$$

From the above equations, it can be noted that

$$\sigma = \sigma_x \quad \text{on plane } \phi = 0°$$

$$\text{and } \sigma = 0 \quad \text{on plane } \phi = 90° \qquad [4.3(a)]$$

Similarly, it can be noted that

$$\tau = 0 \quad \text{on planes } \phi = 0 \text{ and } \phi = 90°$$

$$\text{and } \tau = -\sigma_x/2 \quad \text{on plane } \phi = 45° \qquad [4.3(b)]$$

It can be noted from the above equations that a body subjected to direct stress develops a maximum shear stress equal to half the applied stress on planes inclined at 45° to the direction of applied stress.

The resultant stress on the plane is given by

$$\sigma_R = \sqrt{(\sigma^2 + \tau^2)} = \sigma_x \sqrt{(\cos^4\phi + \sin^2\phi\cos^2\phi)} = \sigma_x \cos\phi \qquad [4.3(c)]$$

Obviously, the resultant stress has its maximum value ($= \sigma_x$, the applied stress) on a plane with $\phi = 0°$ (the plane normal to the direction of applied stress); planes parallel to the direction of applied stress ($\phi = 90°$) has zero stress.

Problem I A square bar of size 20.0 × 20.0 mm is subjected to direct tensile force. Find the largest value of tensile force the bar can sustain, if its shear strength is 100.0 MPa. Determine the stresses on a plane inclined at 60° with the bar axis under the maximum load.

Solution I Maximum shear stress in a bar subjected to direct tensile stress is half the direct stress on a plane inclined at 45° with the member axis (Eqn.4.3(b)). Thus, we have

$$\text{Maximum allowable direct stress} = 2\times \text{ shear strength} = 2 \times 100.0 \text{ MPa} = 200.0 \text{ MPa}$$

Maximum allowable force $=$ max. allowable stress $\times$ area

$= 200.0 \times (20.0 \times 20.0)$ N $= 80.0$ kN

Considering Fig.4.3, let plane EF be inclined at $\phi = 60°$ with the bar axis. If A = area of the bar section, we have

Area of the plane $A' = A/\sin\phi$

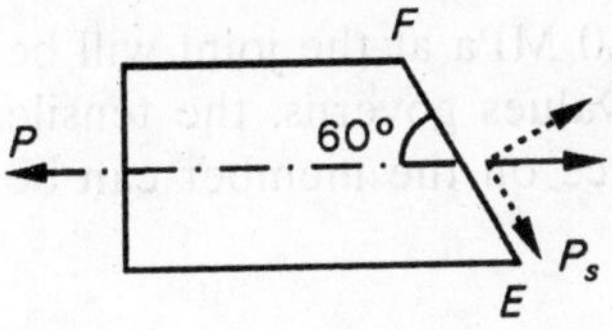

Fig.4.3 Forces on an inclined plane (Problem I)

The direct force P on the section can be resolved in directions normal (P_n) and along (P_s) plane EF. We have

$P_n = P \sin\phi$ and $P_s = P\cos\phi$

Direct stress on plane EF $= P_n/A' = P\sin^2\phi/A$

$= 80\ 000.0 \sin^2(60)/(20.0 \times 20.0)$ MPa $= 150.0$ MPa

Shear stress on plane EF $= P_s/A' = P(\cos\phi)(\sin\phi)/A$

$= 80\ 000.0 \cos(60)\sin(60)/(20.0 \times 20.0)$ MPa $= 86.60$ MPa

Problem II Two timber pieces of section 60.0×100.0 mm are joined with epoxy along a plane inclined at 30° to the X-axis. Find the maximum force the member can sustain if the strength of the joint is 5.0 MPa in tension and 3.0 MPa in shear.

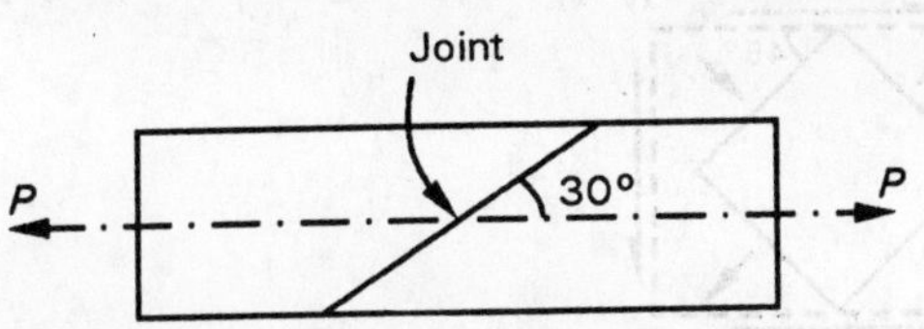

Fig.4.4 Stress on a joint (Problem II)

Solution II Referring to Fig.4.4, the allowable direct tensile stress σ_x can be computed in terms of the allowable stresses developed at the joint; the lower of the corresponding direct stresses will be the maximum allowable value. The joint is inclined at 30° with the X-axis; its inclination with the Y-axis will be 60°. Thus, from Eqn.4.2 we have

Direct stress at the joint $\sigma_a = \sigma_x \cos^2\phi = \sigma_x \cos^2(60) = 0.25\,\sigma_x$

or $\sigma_x = \sigma_a/0.25 = 4.0\sigma_a$

The maximum allowable tensile stress for a tensile strength (σ_a) of 5.0 MPa at the joint will be accordingly $(4.0 \times 5.0) = 20.0$ MPa.

$$\text{Shear stress at the joint} \quad \tau_a = \sigma_x \sin\phi\cos\phi = \sigma_x \sin(60)\cos(60)$$
$$= 0.433\,\sigma_x \quad \text{or} \quad \sigma_x = \tau_a/0.433 = 2.31\tau_a$$

The maximum allowable tensile stress for a shear strength (τ_a) of 3.0 MPa at the joint will be accordingly $(2.31 \times 3.0) = 6.93$ MPa. Since the lower of these two values governs, the tensile strength of the member will be 6.93 MPa. The allowable tensile force on the member can be computed as

$$P = 6.93 \times (60.0 \times 100.0)\ \text{N} = 41.58\ \text{kN}$$

4.3 Pure shear stress

Consider a body under shear stresses shown in Fig.4.5(a). The stresses on a plane inclined at ϕ with face DC of the body are indicated in Fig.4.5(b). The equation of equilibrium of forces in the direction of normal stress can be written as

$$\sigma(EC) - \tau_{xy}\sin\phi(DC) - \tau_{xy}\cos\phi(ED) = 0$$

or $\sigma(l/\cos\phi) - \tau_{xy}\sin\phi(l) - \tau_{xy}\cos\phi(l\tan\phi) = 0$

Simplifying the above equation, we obtain

$$\sigma = \tau_{xy}\sin(2\phi) \qquad [4.4(a)]$$

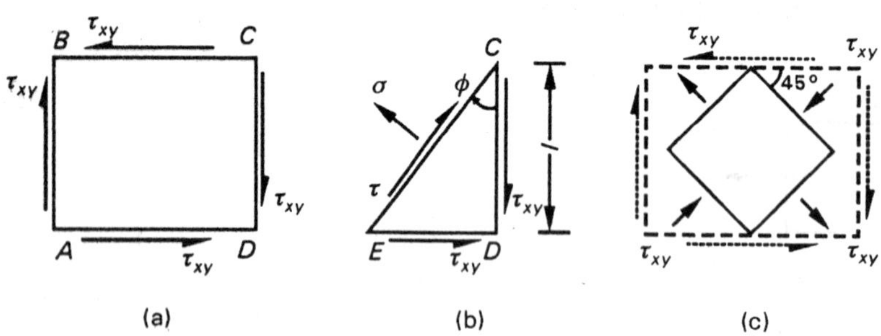

Fig.4.5 An element under pure shear

Similarly, the equation of equilibrium of forces in the direction of shear stress τ can be expressed as

$$\tau(EC) - \tau_{xy}\cos\phi(DC) + \tau_{xy}\sin\phi(ED) = 0$$

or $\tau\,(l/\cos\phi) - \tau_{xy}\cos\phi\,(l) + \tau_{xy}\sin\phi\,(l\tan\phi) = 0$

Simplifying the above equation we obtain

$$\tau = +\tau_{xy}\cos(2\phi) \qquad [4.4(b)]$$

It can be noted from Eqns.4.4(a) and 4.4(b) that

$$\sigma = 0 \text{ and } \tau = \tau_{xy} \text{ on plane } \phi = 0°; \quad \sigma = 0 \text{ and } \tau = -\tau_{xy} \text{ on plane } \phi = 90°$$

$$\sigma = \tau_{xy} \text{ and } \tau = 0 \text{ on plane } \phi = 45°; \quad \sigma = -\tau_{xy} \text{ and } \tau = 0 \text{ on plane } \phi = 135° \qquad [4.5]$$

It can also be noted that the resultant stress σ_R on any plane is equal to the applied shear stress τ_{xy}. Equations 4.5 indicate that planes inclined at 45° are subjected to a tensile stress of magnitude τ_{xy}, while planes inclined at 135° are subjected to a compressive stress of the same magnitude; there is no shear stress on these planes. The stress condition on these planes is indicated in Fig.4.5(c).

4.4 Bi-axial stress condition

Figure 4.6(a) indicates an element of a body under bi-axial stress condition. The stresses in subsequent analyses will be designated by appropriate subscripts as mentioned in Section 4.1. The equilibrium of forces on plane CE inclined at an angle ϕ with face CD is considered. The stresses on segment CDE are indicated in Fig.4.6(b). Equilibrium condition of forces in the direction of σ, the normal stress on plane CE can be developed as

$$\sigma\,(EC) - \sigma_x\cos\phi\,(CD) - \sigma_y\sin\phi\,(ED) = 0$$

$$\sigma\,(l/\cos\phi) - \sigma_x\cos\phi(l) - \sigma_y\sin\phi\,(l\tan\phi) = 0$$

or $$\sigma = \sigma_x\cos^2\phi + \sigma_y\sin^2\phi \qquad [4.6(a)]$$

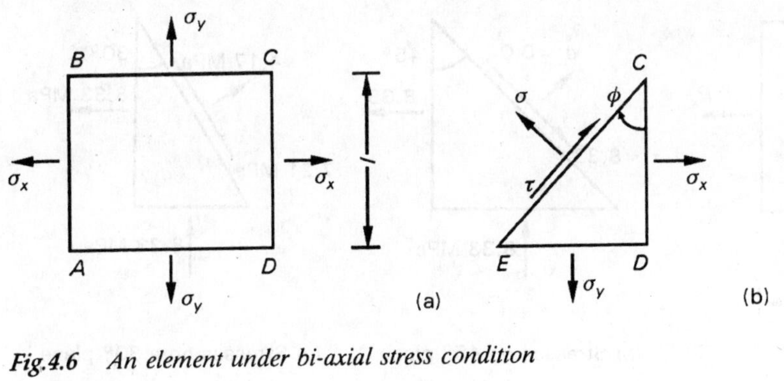

Fig.4.6 An element under bi-axial stress condition

Similarly, resolving the forces in the direction of shear stress τ, the equilibrium condition can be written as

$$\tau\,(EC) + \sigma_x \sin\phi\,(CD) - \sigma_y \cos\phi\,(ED) = 0$$

$$\tau\,(l/\cos\phi) + \sigma_x \sin\phi(l) - \sigma_y \cos\phi\,(l \tan\phi) = 0$$

or $$\tau = -[(\sigma_x - \sigma_y)/2]\sin(2\phi) \qquad [4.6(b)]$$

Equation 4.6(a) indicates that the normal stress on a plane varies from σ_x to σ_y as the angle ϕ is varied from 0° to 90°. Equation 4.6(b) indicates that shear stress is zero on planes with $\phi = 0°$ and 90°; the shear stress has a maximum value of $[(\sigma_x - \sigma_y)/2]$ on planes with $\phi = 45°$ and 135°. It can be noted that when $\sigma_y = 0$, Eqns.4.6(a) and (b) degenerate into Eqn.4.2(a) and (b), respectively.

Problem III A rectangular block of size 120.0 × 150.0 mm and thickness 100.0 mm is subjected to a tensile force of 100.0 kN on the shorter side and a compressive force of 125.0 kN on the longer side. Find the stresses on planes inclined at 45° and 30° with the Y-axis.

Solution III The forces on the block are indicated in Fig.4.7(a). From the given data, we have

$$\sigma_x = +100.0 \times 10^3/(120.0 \times 100.0) = +8.33 \text{ MPa}$$

and $$\sigma_y = -125.0 \times 10^3/(150.0 \times 100.0) = -8.33 \text{ MPa}$$

(i) Stresses on a plane inclined at 45° with the Y-axis From Eqns.4.6, we have

$$\sigma = \sigma_x \cos^2\phi + \sigma_y \sin^2\phi = +8.33\cos^2(45) - 8.33\sin^2(45) = 0.0$$

and $$\tau = -(\sigma_x - \sigma_y)\sin\phi\cos\phi = -(8.33 + 8.33)\sin(45)\cos(45) \text{ MPa} = -8.33 \text{ MPa}.$$

The stress condition is indicated in Fig.4.7(b).

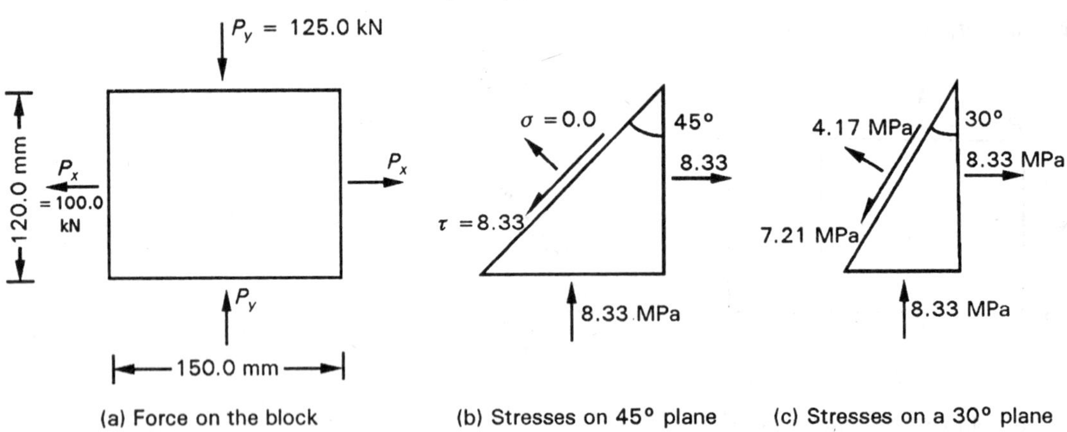

Fig.4.7 Bi-axial loading on a block (Problem III)

(ii) Stresses on a plane inclined at* 30° *with the Y-axis The stresses on the plane can be computed as in the previous case. From Eqns.4.6, we have

$$\sigma = +8.33\cos^2(30) - 8.33\sin^2(30) = +4.17 \text{ MPa}$$

and $\tau = -(8.33 + 8.33)\sin(30)\cos(30) \text{ MPa} = -7.21 \text{ MPa}$

The stress condition is indicated in Fig.4.7(c).

4.5 General stress analysis (combined bi-axial and shear stresses)

The stresses at a point in a body may comprise, in general, stress components σ_x, σ_y and τ_{xy} as indicated in Fig.4.1 (a combination of the cases discussed in Sections 4.3 and 4.4).

Considering a plane CE inclined at an angle ϕ with the Y-axis as indicated in Fig.4.8, the direct stress σ and shear stress τ on the plane can be determined from static equilibrium conditions as in the previous cases. The dimensions of the element are denoted in terms of infinitesimal lengths; let the dimensions be $ED = dx$, $CD = dy$ and $CE = ds$ (Fig.4.8). From the geometry of the body, we have

$$dx = ds\sin\phi \quad \text{and} \quad dy = ds\cos\phi \qquad [4.7]$$

Resolving the forces on segment CDE in the direction of normal stress σ on plane CE, we have

$$\sigma\, ds = \sigma_x\, dy\cos\phi + \sigma_y\, dx\sin\phi + \tau_{xy}\, dy\sin\phi + \tau_{xy}\, dx\cos\phi \qquad [4.8]$$

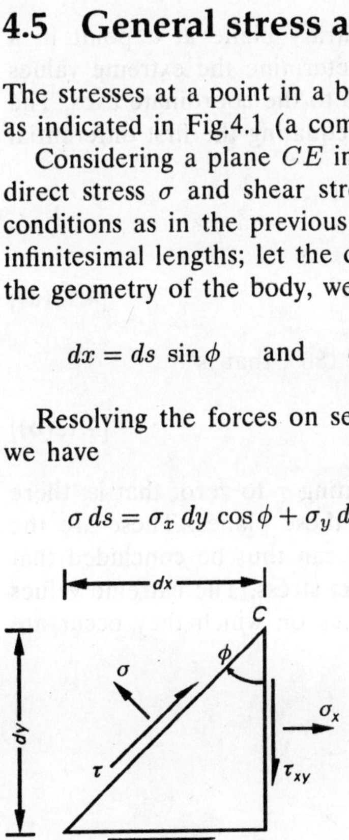

Fig.4.8 Stresses on an elemental plane

Substituting for dx and dy from Eqn.4.7 in Eqn.4.8, and simplifying, we obtain

$$\begin{aligned}\sigma &= \sigma_x\cos^2\phi + \sigma_y\sin^2\phi + 2\,\tau_{xy}\sin\phi\cos\phi \\ &= \{(\sigma_x + \sigma_y)/2\} + \{(\sigma_x - \sigma_y)/2\}\cos 2\phi + \tau_{xy}\sin 2\phi \qquad [4.9]\end{aligned}$$

Similarly, resolving the forces acting on segment CDE in the direction of shear stress τ on plane CE, we have

$$\tau\, ds = -\sigma_x\, dy \sin\phi + \sigma_y\, dx \cos\phi + \tau_{xy}\, dy \cos\phi - \tau_{xy}\, dx \sin\phi \qquad [4.10]$$

Simplifying the above equation, we obtain

$$\begin{aligned} \tau &= -(\sigma_x - \sigma_y) \sin\phi \cos\phi + \tau_{xy} (\cos^2\phi - \sin^2\phi) \\ &= -\{(\sigma_x - \sigma_y)/2\} \sin 2\phi + \tau_{xy} \cos 2\phi \end{aligned} \qquad [4.11]$$

Equations 4.9 and 4.11 help determine stresses on any arbitrary plane at a point in a body under general state of stress. These equations also help determine the extreme values (maximum and minimum) of stresses and their directions relative to the coordinate axes. The extreme values of direct stress on plane CE can be obtained by equating the first differential of Eqn.4.9 with reference to ϕ to zero. We have,

$$[d\sigma/d\phi] = -2\{(\sigma_x - \sigma_y)/2\} \sin 2\phi + \tau_{xy} \cos 2\phi = 0$$

or $\tan 2\phi = 2\tau_{xy}/(\sigma_x - \sigma_y)$ [4.12(a)]

The above equation yields two values of (2ϕ) which differ by 180°; that is

$$\theta_2 = \theta_1 + 90° \qquad [4.12(b)]$$

Equation 4.12(a) can also be obtained from Eqn.4.11 by equating τ to zero; that is, there exist certain values of $\phi = \theta$ such that shear stress vanishes on these planes. These are the same planes on which extreme values of normal stress occur. It can thus be concluded that shear stresses are zero on the planes with extreme values of direct stress. The extreme values of direct stress are known as the *principal stresses*, and the planes on which they occur are known as the *principal planes*.

Further, Eqn.4.12(a) yields

$$\sin 2\theta = \pm 2\tau_{xy} \Big/ \sqrt{[(\sigma_x - \sigma_y)^2 + 4\tau_{xy}^2]}$$

and $\cos 2\theta = \pm(\sigma_x - \sigma_y) \Big/ \sqrt{[(\sigma_x - \sigma_y)^2 + 4\tau_{xy}^2]}$ [4.13]

It should be noted that right hand sides of both the above equations should have either positive or negative sign. Substituting for the values of $\sin 2\theta$ and $\cos 2\theta$ from the above equations in Eqn.4.9 we obtain two values for the direct stress corresponding to the two values of θ given by Eqn.4.12(b). The two values of principal stresses are given by

$$\sigma_{1,2} = \{(\sigma_x + \sigma_y)/2\} \pm \sqrt{[\{(\sigma_x - \sigma_y)/2\}^2 + \tau_{xy}^2]} \qquad [4.14]$$

Principal stresses are referred to as σ_1 and σ_2. The former is the larger of the two, and is termed as the *major principal stress* while the other is called the *minor principal stress*. The corresponding planes are termed major and minor principal planes. In order to obtain the planes of maximum shear stress, Eqn.4.11 should be differentiated with reference to ϕ and equated to zero. This operation yields

$$(d\tau/d\phi) = -\{(\sigma_x - \sigma_y)/2\} 2\cos 2\phi - \tau_{xy} 2 \sin 2\phi = 0$$

or $$\tan 2\phi = -(\sigma_x - \sigma_y)/(2\tau_{xy}) \qquad [4.15]$$

It is easy to note from Eqns.4.12(a) and 4.15 that the planes of maximum direct stress and those of maximum shear stress differ by 45° (since the angles 2θ and 2ϕ differ by 90°). In other words, if ϕ_1' and ϕ_2' are the planes of maximum shear stress

$$\phi_1' = \theta_1 - 45° \quad \text{and} \quad \phi_2' = \theta_2 - 45° \qquad [4.16]$$

It can be shown that

$$\tau_{max} = (\sigma_1 - \sigma_2)/2 \qquad [4.17]$$

That is, maximum shear stress is half the difference between the principal stresses. It can be noted that Eqns.4.9 and 4.11 degenerate into the previous cases (Sections 4.3–4.5) when the appropriate applied stresses are eliminated.

Note The values of principal stresses can be computed from Eqn.4.14 and the inclination of principal planes from Eqn.4.12. However, the principal stresses and planes cannot be correlated directly. It is preferable to compute the angles of principal planes (Eqn.4.12(a)) and determine the corresponding principal stresses from Eqn.4.9. This procedure is adopted in the example problems.

Problem IV Stresses at a point in a body are $\sigma_x = -160.0$ Nmm^{-2}, $\sigma_y = +350.0$ Nmm^{-2} and $\tau_{xy} = -200.0$ Nmm^{-2}. Determine the corresponding principal stresses and the planes, and the stresses on a plane inclined at −30° to the X-axis.

Solution IV *Principal planes* The given stress condition is indicated in Fig.4.9(a). The inclination of principal planes with the Y-axis can be determined from Eqn.4.12(a); substituting the given values, we have

$$\tan(2\theta) = 2\tau_{xy}/(\sigma_x - \sigma_y) = [2(-200.0)/(-160.0 - 350.0)] = +0.7843$$

The above equation yields two values for angle (2θ), namely

$$(2\theta) = 38.11° \quad \text{and} \quad 218.11°$$

$$\theta_1 = +19.1° \quad \text{and} \quad \theta_2 = +109.1°$$

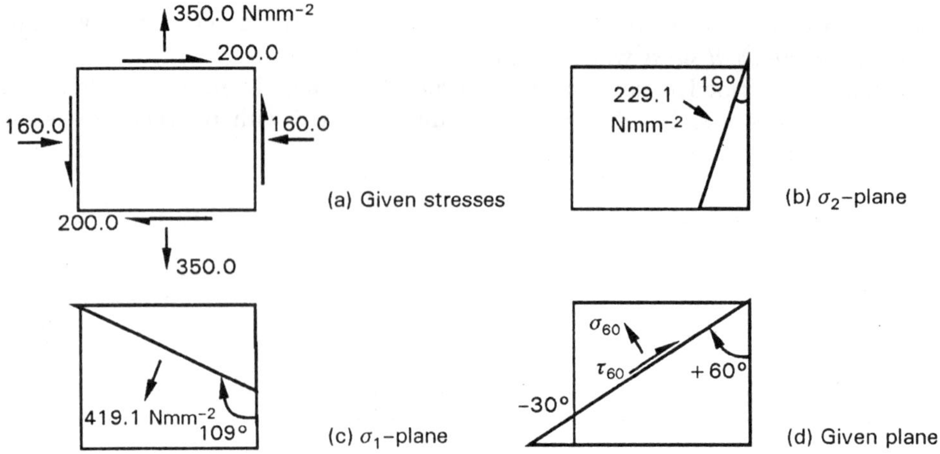

Fig.4.9 Stress analysis (Problem IV)

Principal stresses The principal stresses can be determined from Eqn.4.14. Substituting the given values of stresses, we have

$$\begin{aligned}\sigma_{1,2} &= \{(\sigma_x+\sigma_y)/2\} \pm \sqrt{[\{(\sigma_x-\sigma_y)/2\}^2+\tau_{xy}^2]}\\ &= \{(-160.0+350.0)/2\} \pm \sqrt{[\{-160.0-350.0)/2\}^2+(-200.0)^2]}\\ &= +419.1 \text{ Nmm}^{-2} \quad \text{and} \quad -229.1 \text{ Nmm}^{-2}\end{aligned}$$

However, the principal planes and stresses cannot be correlated from these computations. It is preferable to compute the values of principal stresses from Eqn.4.9(a); the subscript of the stress indicates the plane of the stress in the computations. We have,

$$\begin{aligned}\sigma_{19.1} &= \sigma_x \cos^2\phi + \sigma_y \sin^2\phi + 2\,\tau_{xy} \sin\phi \cos\phi\\ &= -160.0\cos^2(19.1) + 350.0\sin^2(19.1) + 2(-200.0)\sin(19.1)\cos(19.1)\\ &= -229.1 \text{ MPa}\end{aligned}$$

Similarly, we have

$$\begin{aligned}\sigma_{109.1} &= -160.0\cos^2(109.1) + 350.0\sin^2(109.1) + 2(-200.0)\sin(109.1)\cos(109.1)\\ &= +419.1 \text{ MPa}\end{aligned}$$

From the convention adopted, the major principal stress $\sigma_1 = +419.1$ MPa and occurs on the plane inclined at 109.1° with the Y-axis; the minor principal stress $\sigma_2 = -229.1$ MPa and occurs on the plane inclined at 19.1°. The corresponding planes are indicated in Fig.4.9(b) and (c).

Stresses on the given plane The given plane is inclined at −30° with the X-axis or at +60° with the Y-axis (Fig.4.8(d)). Substituting the given values in Eqn.4.9, we have

$$\begin{aligned}\sigma_{60} &= \sigma_x \cos^2\phi + \sigma_y \sin^2\phi + 2\tau_{xy}\sin\phi\cos\phi \\ &= [(-160.0)\cos^2(60) + (350.0)\sin^2(60) + 2(-200.0)\cos(60)\sin(60)] \text{ MPa} \\ &= +49.3 \text{ MPa}\end{aligned}$$

The shear stress on the given plane can be computed from Eqn.4.11; we have

$$\begin{aligned}\tau_{60} &= [-\{\sigma_x - \sigma_y)/2\}\sin(2\phi) + \tau_{xy}\cos(2\phi)] \\ &= -[\{-160.0 - 350.0)/2\}\sin(120) + (-200.0)\cos(120)] = +320.8 \text{ MPa}.\end{aligned}$$

The stresses are indicated in Fig.4.9(d).

4.6 Relation between Young's modulus and rigidity modulus

It was shown in Section 4.3 that when a body is subjected to pure shear, direct stresses of the same magnitude (as the applied shear stress) occur on planes inclined at 45° and 135° with the axis of reference (Fig.4.5 and Eqn.4.5). Considering a square element $ABCD$ of sides l shown in Fig.4.10, diagonal BD is subjected to tensile stress and diagonal AC to compressive stress. Let $AB'C'D$ denote the body deformed under the applied stress. It can be noted that, because of the deformation of the body, the right angles BAD and BCD increase while the right angles ABC and ADC decrease. With the usual notation, we have (Eqn.1.25)

Linear strain in BD, $\varepsilon = \tau(1+\nu)/E$ $[\sigma = \tau]$

Linear strain in AC, $\varepsilon' = -\tau(1+\nu)/E = -\varepsilon$ [4.18(a)]

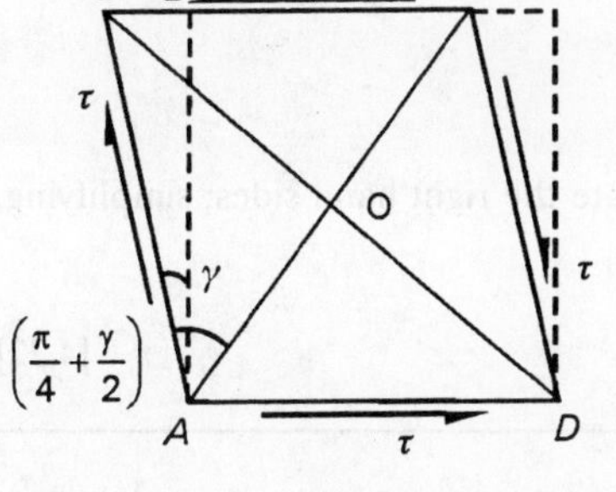

Fig.4.10 Pure shear on a square element

Using the above expressions, the changed lengths of the diagonals can be expressed as

$B'D = \sqrt{2}\,l(1+\varepsilon)$

and $C'A = \sqrt{2}\, l\,(1 - \varepsilon)$ [4.18(b)]

The shear strain in the element by definition denotes the change in the right angles of the body; shear strain in the body is given by

$$\gamma = \tau / G \quad [4.19]$$

It can be assumed that half the angular change occurs in angle $B'AC'$ and the other half in angle $C'AD$. Thus, angle $B'AO$ of the deformed body is given by $[(\pi/4) + (\gamma/2)]$. Noting that O bisects the diagonals AC' and $B'D$, we have from Eqn.4.18 (b)

$$\tan[(\pi/4) + (\gamma/2)] = B'O/AO = (1 + \varepsilon)/(1 - \varepsilon) \quad [4.20(a)]$$

The left hand side of the above equation can be simplified, after noting that $\tan(\gamma/2) \simeq (\gamma/2)$, as

$$\begin{aligned} \tan[(\pi/4) + (\gamma/2)] &= [\tan(\pi/4) + \tan(\gamma/2)]\ /[\tan(\pi/4) - \tan(\gamma/2)] \\ &= [1 + (\gamma/2)]/[1 - (\gamma/2)] \end{aligned} \quad [4.20(b)]$$

Comparing Eqns.4.20(a) and 4.20(b), we have

$$\varepsilon = (\gamma/2)$$

Substituting for ε and γ from Eqns.4.18(a) and 4.19 in the above equation, we obtain

$$G = E/[2(1 + \nu)] \quad [4.21]$$

Note The above equation can be combined with the relation between K and E developed in Section 1.12. Equation 4.21 can be re-written as

$$\nu = [\{E/(2G)\} - 1]$$

Similarly from Eqn.1.28, we have

$$\nu = \frac{1 - E/(3K)}{2}$$

Since both the equations pertain to the same material, we can equate the right hand sides; simplifying, we obtain

$$E = \frac{9GK}{(G + 3K)} \quad [4.22]$$

4.7 Mohr's stress circle

Mohr observed that Eqns.4.9 and 4.11 can be expressed in graphical form by taking σ and τ as co-ordinates. Direct stresses are considered positive to the right of the origin and shear stresses are positive above the origin. There exists a corresponding point for stresses on each plane on this diagram; the co-ordinates of the point represent the values of σ and τ for the plane. Figure 4.11 gives a graphical representation of the stresses.

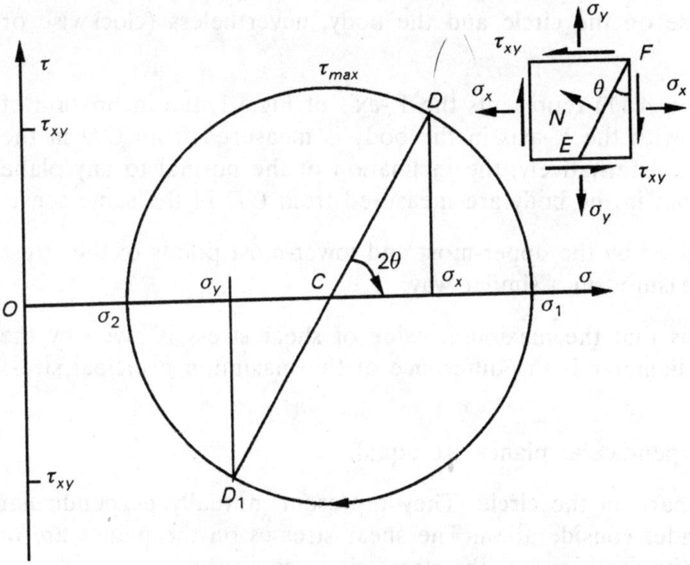

Fig.4.11 Mohr's stress circle

The circle shown in Fig.4.11 is constructed by noting the stress conditions on vertical faces as point D and those on horizontal faces as point D_1 on the stress axes, and drawing a circle to pass through them. The convention here being that the shear stresses on faces AB and DC form a clockwise couple and hence are considered positive, while those on faces BC and AD form a counter-clockwise couple, and hence are negative (Fig.4.1). The stresses on the body are also indicated above the circle in Fig.4.11. It is easy to note that the centre of the circle is the point $\{(\sigma_x + \sigma_y)/2\}$, and the

$$\text{radius of the circle} = [\{(\sigma_x - \sigma_y)/2\}^2 + \tau_{xy}^2]^{0.5}$$

It can be seen that the principal stresses given by Eqn.4.14 are represented by σ_1 and σ_2 indicated on the intercepts made by the circle on the σ-axis. Further, point D on the circle denotes the stresses on face CD of Fig.4.1(b) and point D_1 denotes those on face DE of the figure. Principal plane is inclined to DD_1 at an angle 2θ (measured clockwise) given by Eqn.4.12. The inclination of line DD_1 with the σ-axis is measured in the same sense as that in Fig.4.1(b). This diagram, known as Mohr's circle, brings out several features in this regard.

(i) As the face DC (Fig.4.1(b)) is rotated clockwise about C, the corresponding plane in Mohr's circle is generated by rotating CD about C such that the planes $0 \leq \phi \leq 90°$ are represented by the points on the semicircle $D\sigma_1 D_1$ (Fig.4.11), and those for which $90° < \phi < 180°$ are represented by the semicircle $D_1\sigma_2 D$.

In other words, the Y-axis of the body is represented by CD and the X-axis by CD_1 on the stress diagram of Fig.4.11.

(ii) The inclination of the principal planes is given by half the angle measured from CD to the pertinent stress (θ_1, if measured towards σ_1, and θ_2 towards σ_2). The angle should be measured in the same sense on the circle and the body, nevertheless (clockwise or counter-clockwise).

Since CD on the stress diagram (Fig.4.11) represents the Y-axis of Fig.4.1, the inclination of any plane (face CE in Fig.4.1(b)) with the Y-axis in the body is measured from CD in the same direction as that in the body. Alternatively, the inclination of the normal to any plane from the X-axis (Fig.4.1(b)) and that in the body are measured from CD in the same sense.

(iii) Maximum shear stresses are given by the upper-most and lower-most points of the stress circle. Their directions are measured in a similar way.

Referring to Eqn.4.17, it is obvious that the maximum value of shear stress is given by the radius of the stress circle, whose diameter is the difference of the maximum principal stress values.

(iv) Shear stresses on mutually perpendicular planes are equal.

The points D and D_1 are 180° apart on the circle. They represent mutually perpendicular planes in the body at the point under consideration. The shear stresses on the planes are of the same magnitude (but of opposite sense as per the stress circle convention).

(v) The sum of direct stresses on any set of mutually perpendicular planes remains the same.

The sum of direct stresses at D and D_1 and that of stresses σ_1 and σ_2 (on mutually perpendicular planes) are equal. This can be verified for the values of direct stresses at the ends of any diameter of the stress circle. The representation of stresses on Mohr's circle for a few typical stress conditions is indicated in Fig.4.12.

4.7.1 Negative shear stresses

Referring to Fig.4.12(a), the shear stress is negative according to the convention adopted (shear stresses on vertical faces form counter-clockwise couple). Thus, the shear stresses are marked on the negative side of the τ-axis for the vertical faces (D), and on positive side for the horizontal faces (D_1). Accordingly, 2θ on the circle is measured in the counter-clockwise (negative) direction from CD. The corresponding plane on the body will be inclined at $-\theta$ with the vertical face (or the normal on the principal plane will be inclined at $-\theta$ with the X-axis) as indicated in the figure.

4.7.2 $\sigma_x < \sigma_y$

The direct stress on vertical faces σ_x was assumed to be greater than σ_y in the previous figures. If $\sigma_x < \sigma_y$, the stress circle can be represented as indicated in Fig.4.12(b). The shear stress is positive in this case and is marked in the positive direction from σ_x. The angle θ is still measured from CD to the σ-axis as indicated in the figure; the corresponding plane in the body is also indicated.

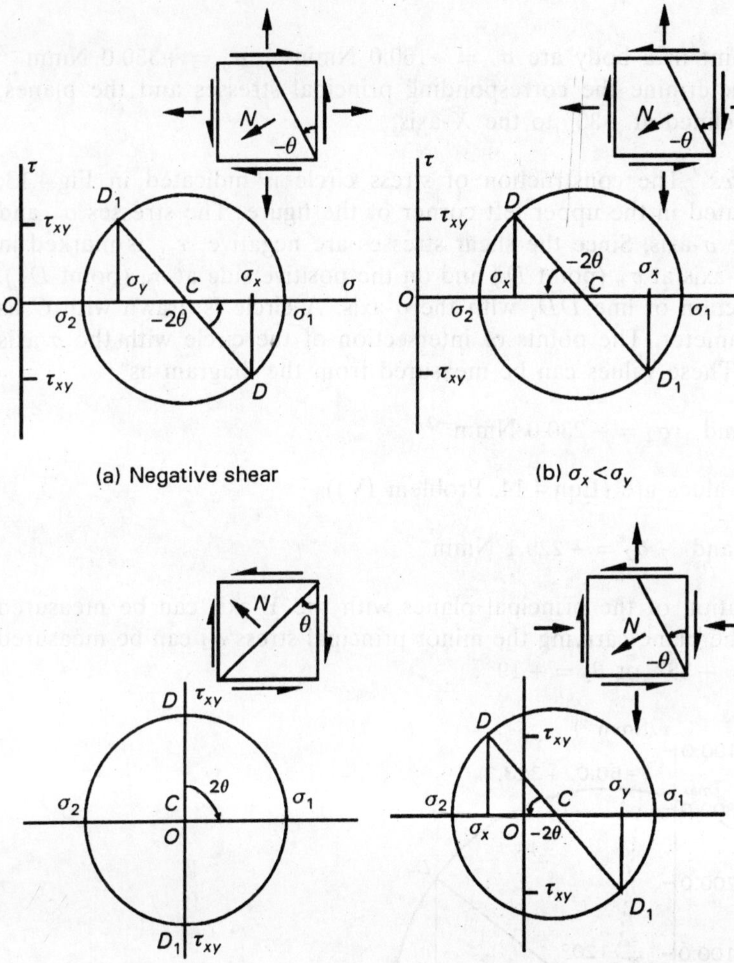

Fig.4.12 A few typical stress circles

4.7.3 $\sigma_x = \sigma_y = 0$

The corresponding stress circle is shown in Fig.4.12(c). In this case, D on the stress diagram lies on the positive side of the τ-axis (positive shear on vertical faces), while D_1 denotes the stress on horizontal faces (negative shear). The centre of the stress circle lies at the origin, and the inclination of the major principal planes is 90° on the circle (or 45° in the body).

4.7.4 $\sigma_x < 0$

This is similar to the case $\sigma_x < \sigma_y$ (Fig.4.12(b)). The corresponding stress circle is indicated in Fig.4.12(d).

Problem V Stresses at a point in a body are $\sigma_x = -160.0$ Nmm^{-2}, $\sigma_y = +350.0$ Nmm^{-2} and $\tau_{xy} = -200.0$ Nmm^{-2}. Determine the corresponding principal stresses and the planes, and the stresses on a plane inclined at $-30°$ to the X-axis.

Solution V *Principal stresses* The construction of stress circle is indicated in Fig.4.13; stresses on the body are indicated in the upper left corner of the figure. The stresses σ_x and σ_y are marked out first on the σ-axis. Since the shear stresses are negative, τ_{xy} is marked in the negative direction of the τ-axis at σ_x (point D) and on the positive side at σ_y (point D_1). Let C be the point of intersection of line DD_1 with the σ-axis. A circle is drawn with C as the centre and DD_1 as its diameter. The points of intersection of the circle with the σ-axis give the values of σ_1 and σ_2. These values can be measured from the diagram as

$$\sigma_1 = +420.0 \text{ Nmm}^{-2} \quad \text{and} \quad \sigma_2 = -230.0 \text{ Nmm}^{-2}$$

The corresponding analytical values are (Eqn.4.14, Problem IV))

$$\sigma_1 = +419.1 \text{ Nmm}^{-2} \quad \text{and} \quad \sigma_2 = -229.1 \text{ Nmm}^{-2}$$

Principal planes The inclination of the principal planes with the Y-axis can be measured from CD. The inclination of the plane carrying the minor principal stress σ_2 can be measured (clockwise from CD) as $2\theta_2 = +38°$ or $\theta_2 = +19°$.

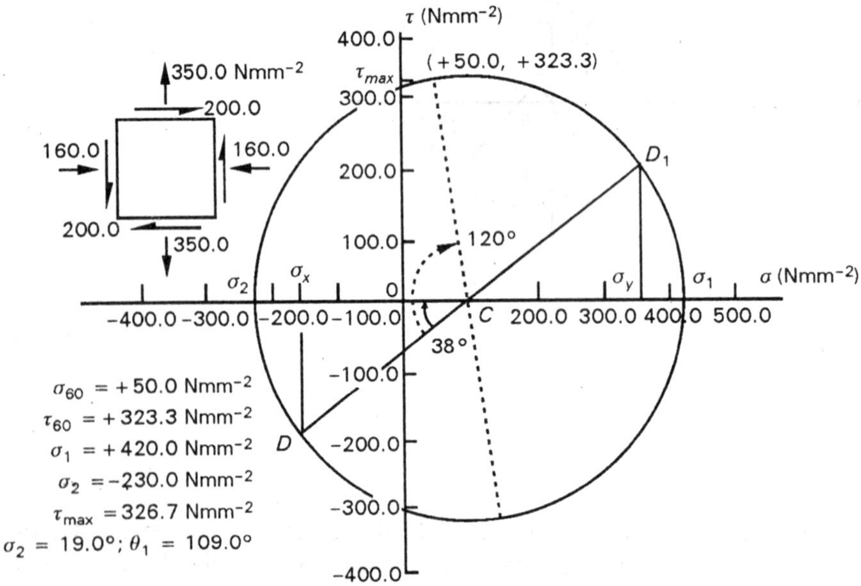

Fig.4.13 Stress analysis using Mohr's circle (Problem V)

The angle indicated in the figure pertains to the plane carrying the stress σ_2, since it is measured towards it. The plane on which σ_2 acts is indicated in Fig.4.9(b). The other angle measured clockwise from CD can be noted to be

$$2\theta_1 = +218^\circ \quad \text{or} \quad \theta_1 = +109^\circ$$

The value of θ_1 pertains to the plane of σ_1. The plane is marked out in Fig.4.9(c). The corresponding analytical values of principal planes are (Eqn.4.12)

$$\theta_1 = +109.1^\circ \quad \text{and} \quad \theta_2 = +19.1^\circ$$

Stresses on the given plane The given plane is inclined at $-30°$ with the X-axis or at $+60°$ with the Y-axis. (That is, its outward normal is inclined at $+60°$ with the X-axis as indicated in Fig.4.1(b)). The corresponding angle on the stress circle will be 120° from CD measured clockwise. The plane is marked in discontinuous lines on the circle (Fig.4.13). The stresses measured from the diagram are

$$\sigma_{60} = +50.0 \text{ Nmm}^{-2} \quad \text{and} \quad \sigma_{60} = +323.3 \text{ Nmm}^{-2}$$

The corresponding analytical values are $+49.3$ Nmm^{-2} and $+320.8$ Nmm^{-2}, respectively (Eqns.4.9 and 4.11). It can be noted that the graphical construction yields values of adequate accuracy.

4.8 Summary

The concepts of two-dimensional (plane) stress analysis are discussed in this chapter. Theoretical analysis as well as the graphical analogy are explained. The application of Mohr's circle to stress analysis is illustrated with the help of a few typical examples. It is also brought out that while theoretical analysis provides a solution to the problem, graphical analysis provides vivid concepts about the state of stress in a body as well.

In addition to the stresses on any desired plane, Mohr's circle yields the magnitudes of principal stresses and the orientation of principal planes. Principal stresses in a body are given by

$$\sigma_{1,2} = \{\sigma_x - \sigma_y)/2\} \pm [\{(\sigma_x - \sigma_y)/2\}^2 + \tau_{xy}^2]^{0.5}$$

Maximum shear stress is given by

$$\tau_{\max} = [\{(\sigma_x - \sigma_y)/2\}^2 + \tau_{xy}^2]^{0.5} = (\sigma_1 - \sigma_2)/2$$

4.9 Exercises

4.1 Show that a body under pure shear is subjected to tensile and compressive stresses as well.

4.2 Show that the maximum shear stress in a body subjected to a direct stress is half the value of the applied stress.

4.3 Develop the expression from basic principles for stresses on an arbitrary plane in a body subjected to a general bi-axial stress condition.

4.4 Show, from basic principles, that shear stress in a body subjected to equal orthogonal stresses is zero.

4.5 Starting from the basic principles, show that a state of pure shear exists in a body subjected to equal stresses of different nature in orthogonal directions.

4.6 Explain why the planes of maximum shear stress and principal stress planes are inclined at 45° to each other.

4.7 Discuss the concepts of Mohr's stress circle in two dimensions.

4.8 Derive the relation between various elasticity modulii of a material.

4.9 Develop the expressions for principal stresses and planes for a body subjected to a bi-axial stress condition.

4.10 Find the principal stresses and their planes in a body for the stress conditions indicated below; $\tau_{xy} = +200.0$ Nmm^{-2} in all the cases.
(i) $\sigma_x = +360.0$ Nmm^{-2}, $\sigma_y = +250.0$ Nmm^{-2}
(ii) $\sigma_x = -360.0$ Nmm^{-2}, $\sigma_y = +250.0$ Nmm^{-2}
(iii) $\sigma_x = -360.0$ Nmm^{-2}, $\sigma_y = -250.0$ Nmm^{-2}
(iv) $\sigma_x = +360.0$ Nmm^{-2}, $\sigma_y = -250.0$ Nmm^{-2}
Determine the stresses on the planes inclined at +45° and +60° to the X-axis for all the cases. Verify these values graphically.

4.11 Determine the maximum stresses in the body for the stress condition $\sigma_x = +80.0$ Nmm^{-2}, $\sigma_y = -100.0$ Nmm^{-2} and $\tau_{xy} = -150.0$ Nmm^{-2}. Indicate their planes, and the stresses on the planes inclined at + 30° and − 60° with the Y-axis.

4.12 What is the maximum permissible value of shear stress on the body in combination with the stress condition $\sigma_x = +160.0$ Nmm^{-2} and $\sigma_y = -80.0$ Nmm^{-2}, if on any plane (i) the direct stress is not to exceed 220.0 Nmm^{-2} and (ii) shear stress is not to exceed 170.0 Nmm^{-2} ?

4.13 Develop Mohr's circle for the stress condition $\sigma_x = +300.0$ Nmm^{-2}, $\sigma_y = -400.0$ Nmm^{-2}, $\sigma_z = +100.0$ Nmm^{-2} and $\tau_{xy} = -250.0$ Nmm^{-2}. Determine the magnitudes and planes of principal stresses and maximum shear stresses.

4.14 If the stresses at a point in a body are $\sigma_x = -100.0$ Nmm^{-2}, $\sigma_y = +150.0$ Nmm^{-2}, and $\tau_{xy} = +150.0$ Nmm^{-2}, determine the maximum shear stress in the body. Estimate the variation in the principal stresses if all the direct stresses are
(i) tensile and (ii) compressive in nature.

5 Analysis of beams

5.0 Introduction

A beam is a structural element subjected to loads transverse to its axis. It is generally assumed that beams are free from axial force (A.F.) or that the influence of A.F. is negligible in comparison to bending effects. Thus, beams are subjected only to bending (flexure), and deflect transverse to their axes. In the absence of A.F., the equation of equilibrium pertaining to horizontal forces (Eqn.I.1) may be ignored, and only two equations of equilibrium are considered in the analysis.

Analysis of beams involves the determination of forces acting on any section of the beam (B.M. and S.F.) as well as the deflections at various sections. Typical beams and their analysis are discussed in this chapter. The concepts of B.M. and S.F. are explained, and their graphical presentation is described.

5.1 Types of beams

Various features of beams were discussed in Chapter I. Beams may be supported in several ways (Fig.I.3), and are classified accordingly. Beams may be single span or multi-span (more than one span) structures.

A beam with both its ends on simple supports is known as a *simply supported beam* (one vertical reaction at each support); a simply supported beam is also referred to sometimes as a *simple beam* (Fig.5.1(a)). It may be recalled that a beam is represented symbolically as a line.

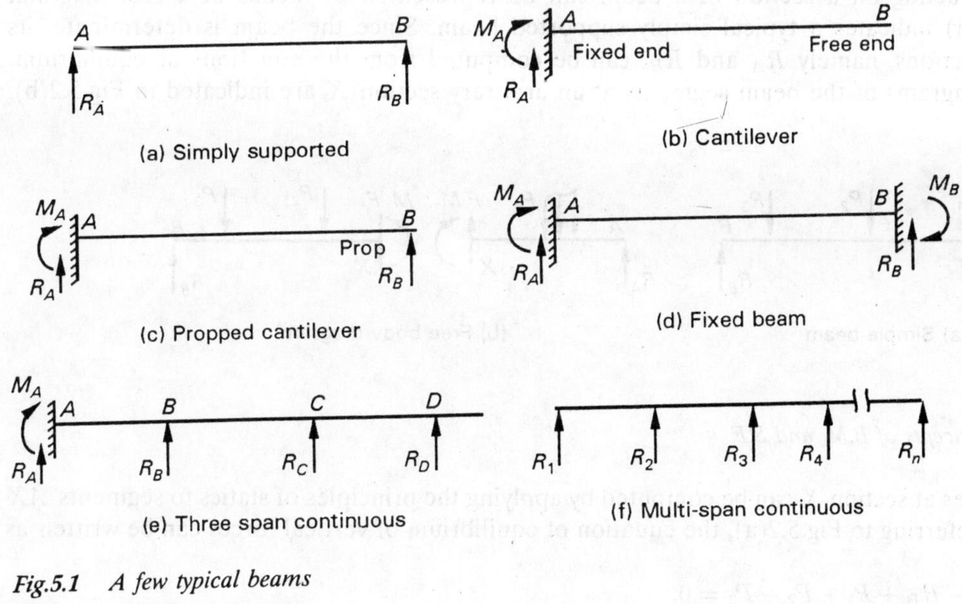

Fig.5.1 A few typical beams

A beam with one end fixed (a vertical reaction and moment) and the other end free (not supported, and hence without any support reactions) is called a *cantilever* (Fig.5.1(b)). Such beams are *statically determinate*, since they have only two unknown support reactions which can be determined from the equations of static equilibrium (Eqn.I.1). Statically determinate beams have the minimum number of reactive forces required for the stability of the structure.

Beams with one end fixed and the other end simply supported (*propped cantilever*; Fig.5.1(c)) have three reaction components, while those with both ends fixed (*fixed beam*; Fig.5.1(d)) have four reaction components. Beams that are supported at more than two sections are known as *continuous beams* (Figs.5.1(e) and (f)). Continuous beams also have more than two unknown reaction compone..ts, depending upon the support conditions, as indicated in Figs.5.1(e) and (f). Beams with more than two reaction components cannot be analysed using the equations of static equilibrium alone and are known as *statically indeterminate* (or *indeterminate*) beams. The number of unknown reaction components in excess of the equations available is known as the *degree of redundancy* (D.O.R.) of the structure.

5.2 Loading on beams

Beams are usually subjected to loads transverse (normal) to their axes. The loading may be distributed (u.d.l. or varying) or a set of concentrated loads. Some of the possible loadings on beams are indicated in Fig.I.1 (Chapter I). Only static loads are considered in the analyses presented here.

5.3 Forces on a beam section

The forces acting on a section of a beam can be represented by means of a F.B. diagram. Figure 5.2(a) indicates a typical simply supported beam. Since the beam is determinate, its support reactions, namely R_A and R_B, can be computed from the equations of equilibrium. The F.B. diagrams of the beam segments at an arbitrary section X are indicated in Fig.5.2(b).

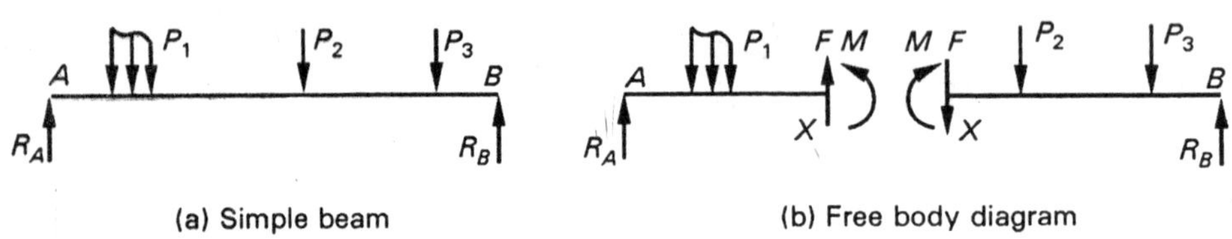

Fig.5.2 Concepts of B.M. and S.F.

The forces at section X can be computed by applying the principles of statics to segments AX and XB; referring to Fig.5.2(a), the equation of equilibrium of vertical forces can be written as

$$-R_A - R_B + P_1 + P_2 + P_3 = 0$$

or $\quad R_A + R_B = P_1 + P_2 + P_3 \quad [5.1]$

Assuming that $P_1 > R_A$, segment AX requires an upward force F of magnitude $(P_1 - R_A)$ at section X to maintain the vertical forces on the body in equilibrium. Further, the force system on segment AX will require a moment M at X for equilibrium of the body.

Since $F = (P_1 - R_A)$, it follows from Eqn.5.1 that $F = [R_B - (P_2 + P_3)]$; segment XB thus requires a downward force F at section X to maintain equilibrium of vertical forces on the body. Similarly, segment XB of the beam will require moment M at section X for equilibrium of the segment. The nature of these forces will be opposite to that of the forces in segment AX.

5.3.1 Shear force and B.M.

Force F represents the unbalanced vertical force on one side (to the left or right) of section X, and is known as the *shear force* (S.F.) at section X.

Shear force at a section can be defined as the sum of all the normal forces on one side of the section. Students should note the difference between force at a section, and S.F. at a section; the former is an absolute quantity, while the latter is the sum of all the forces on one side of a section.

The parameter M at section X of segments AX and XB is known as the *bending moment* (B.M.) at the section. It denotes the sum of moments of all the forces on one side (left or right) of the section.

Bending moment at a section can be defined as the sum of the moments of all the forces on one side of the section. Students should distinguish between moment and B.M. at a section. The former is the sum of the moment of all the forces on the body about the section (and hence equal to zero, when the body is in equilibrium).

5.3.2 Convention for S.F. and B.M.

Figure 5.3 indicates S.F. at a section in the adjacent segments of a beam due to applied loads. The pair of forces in Fig.5.3(a) forms a clockwise couple, and is known as positive S.F. The pair in Fig.5.3(b) forms an anti-clockwise couple, and is thus known as negative S.F. The symbolic notation of the forces is also indicated in the figure.

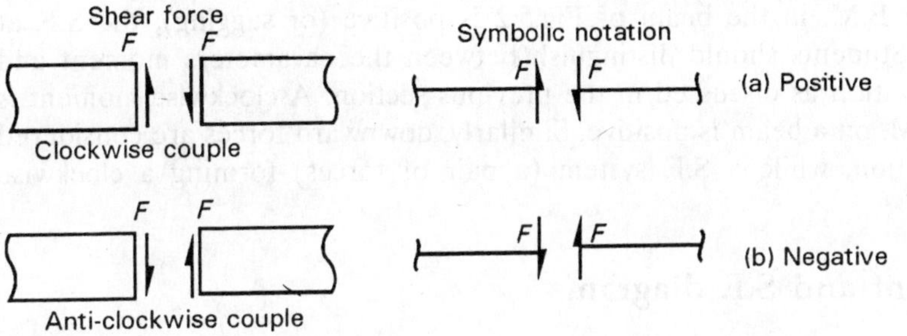

Fig.5.3 Notation for S.F.

Bending moment applied to a beam tends to deflect it normal to its axis; the beam may deflect downward or upward (relative to its original axis) depending upon the nature of the B.M. The moments at the ends of a beam segment are indicated in Fig.5.4. The moments can be replaced by a pair of equal and opposite forces on each face of the beam segment as shown.

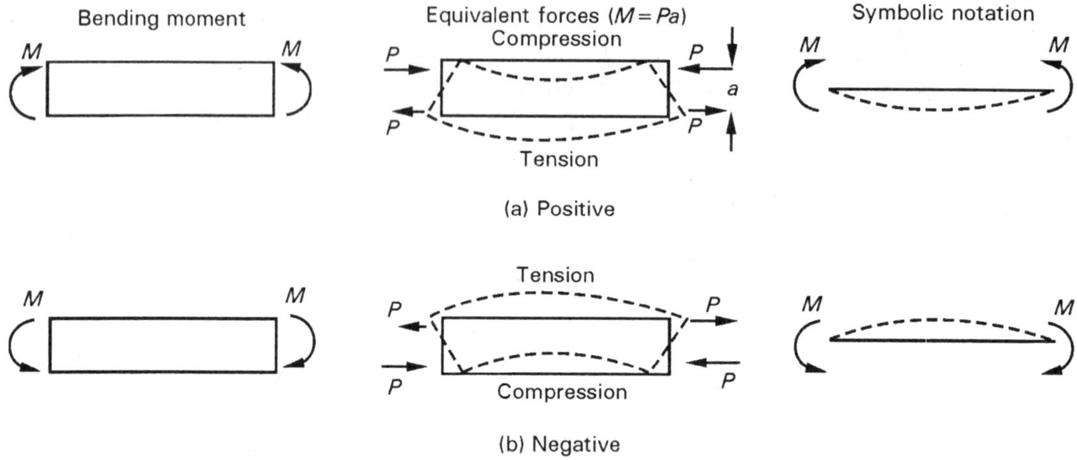

Fig.5.4 Notation for B.M.

It can be seen that the forces shown in Fig.5.4(a) tend to pull the soffit of the beam, while the upper part is compressed. In other words, the moments shown in Fig.5.4(a) cause tension at the soffit of the beam and compression in its upper part. Such a moment induces downward deflection in the beam as shown in discontinuous lines in the figure, and is known as *positive bending moment.* Since the moment causes the beam to sag (deflect downwards, like a suspended rope), it is also known as *sagging bending moment.* A beam is usually denoted symbolically by a line as shown in the figure.

The moments shown in Fig.5.4(b) cause compression on the beam soffit in the direction of the beam axis, while the upper part is in tension. Such a B.M. is referred to as *negative bending moment.* The deflected profile of the segment is also indicated in the figure in discontinuous lines. The beam hogs up (deflects upwards) under such a moment, and is thus known as *hogging bending moment* as well. Often, the term bending is omitted, and it is referred to as moment at a section.

It can be noted that the B.M. in the beam of Fig.5.2 is positive (or sagging); the S.F. at section X is also positive. Students should distinguish between the parameters moment and B.M. while applying the notation as discussed in the previous section. A clockwise moment is positive, while a sagging B.M. on a beam is positive. Similarly, downward forces are considered positive as per the convention, while a S.F. system (a pair of forces) forming a clockwise couple is positive.

5.3.3 Bending moment and S.F. diagrams

Bending moment and S.F. at any section of a beam can be computed from the concepts discussed above. These values can be plotted at each section along the span of the beam. The graphical representation of B.M. along the axis of a structural member is known as *bending moment diagram* (B.M. diagram). Similarly, the graphical representation of S.F. along the axis of a structural member is known as *shear force diagram* (S.F. diagram). It should be noted that B.M. and S.F. diagrams refer to a particular loading system on a beam, which must be specified along with the diagrams.

Mathematical expressions are developed for B.M. and S.F. at a section for the given loading. The values of the parameters are computed at salient sections of the beam, and are indicated graphically. Salient sections of a beam are generally the locations of loads, supports and the mid-span section. The B.M. are plotted on the tension side of a member; since the beam axis is generally horizontal, the positive moments are plotted below the reference line (beam axis), while negative moments are plotted above the reference line.

Positive S.F. is drawn below the reference line, while negative S.F. is shown above the reference line. The B.M. and S.F. diagrams for a few typical beams are indicated in the examples that follow. Subscript x for B.M. and S.F. refers to section X.

5.4 Relation between load, S.F. and B.M.

Bending moment and S.F. in a beam are not independent entities. Both the parameters are functions of loading, span and support conditions of a beam. The relationship between B.M. and S.F. in a beam can be obtained mathematically considering a beam segment AB shown in Fig.5.5. The infinitesimal beam segment of length dx, shown in Fig.5.5, is subjected to varying B.M. and S.F. Let the B.M. be M at the left end of the segment changing to $(M + dM)$ at the right end; let the S.F. change from F at the left end to $(F + dF)$ at the right end. In addition, let the beam segment carry a load of w per unit length.

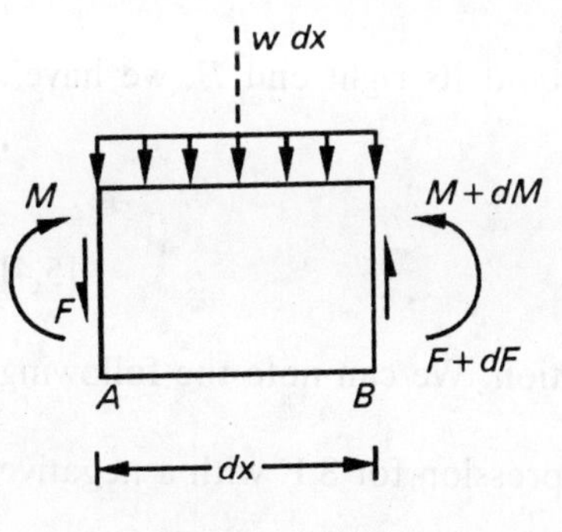

Fig.5.5 Relation between load, S.F. and B.M.

5.4.1 Relation between load and S.F.

Considering the equilibrium of vertical forces on the segment, we have

$$F + w\,dx - (F + dF) = 0$$

or $dF/dx = w$ [5.2]

The above equation can be interpreted in several ways.

1. The first differential of the expression for S.F. with reference to x (length of the beam segment) yields the expression for load on the beam.
2. The slope of the S.F. diagram at a section is the loading at that section.
3. The integral of the loading function over a beam segment yields the difference between the S.F. at the ends of that segment.

   Considering beam segment AB (Fig.5.5), from Eqn.5.2 we have

$$dF = w\,dx$$

Integrating the above equation, we obtain

$$[F_B - F_A] = \int_A^B w\,dx \qquad [5.3]$$

4. The S.F. diagram will have zero slope (or a constant value), if a beam segment is not loaded.

   If a beam segment is not loaded, then we have $w = 0$ $(= dF/dx)$; that is, the S.F. is constant over the segment.
5. The S.F. diagram will have a constant slope over a beam segment, if the segment carries a u.d.l.

   If a beam segment has constant loading (u.d.l.), then w = constant or the the slope of the S.F. diagram over the segment has a constant value (the S.F. varies linearly).

5.4.2 Relation between S.F. and B.M.

Taking the moments of all the forces acting on the segment about its right end B, we have

$$M - F\,dx - w\,dx\,(dx/2) - (M + dM) = 0$$

or $F = -\,dM/dx$ [5.4]

The product of small quantities is neglected in the above equation. We can note the following features from Eqn.5.4.

1. The first differential of the expression for B.M. yields the expression for S.F. with a negative sign.
2. The negative value of the slope of the B.M. diagram at a section is the S.F. at the section.
3. The integral of the S.F. diagram over a segment yields the difference between the B.M. at the ends of the segment.

   Considering beam segment AB (Fig.5.5), from Eqn.5.4, we have

$$dM = -F\,dx$$

Integrating the above equation, we obtain

$$[M_A - M_B] = \int_A^B F\,dx \qquad [5.5]$$

4. At sections with zero S.F., the B.M. diagram has zero slope.
   In other words, the extreme values of B.M. (maximum or minimum) occur at the section where S.F. is zero or changes sign (this feature of B.M. and S.F. diagrams is utilised to determine the extreme values of B.M. in a beam).
5. The constant slope of the B.M. diagram over a beam segment indicates constant S.F. over the segment.

If S.F. over a beam segment is constant, the slope of the B.M. diagram $(-dM/dx)$ will be constant over the segment. That is, the B.M. varies linearly over the segment. These observations help develop S.F. and B.M. diagrams without elaborate computations as will be explained later.

Problem I Draw the B.M. and S.F. diagrams for a simple beam with a concentrated load shown in Fig.5.6.

Solution I The beam shown in Fig.5.6(a) has a concentrated load at a distance a from its left support, and b from the right support $(l = a + b)$. Taking moments of all the forces about support B, we have

$$R_A l - P b = 0 \quad \text{or} \quad R_A = P b/l, \text{ and}$$

$$R_B = P a/l \qquad [\text{since } R_A + R_B = P]$$

Taking section X at a distance x from support A, such that $x \leq a$, the forces at the section can be denoted by F_x and M_x as indicated in Fig.5.6(b). Considering the equilibrium of vertical forces on segment AX, we have

$$-R_A - F_x = 0 \quad \text{or} \quad F_x = -R_A \qquad [\text{I.1(a)}]$$

Taking moments of the forces on segment AX about X, we obtain

$$R_A\, x - M_x = 0 \quad \text{or} \quad M_x = R_A\, x \qquad [\text{I.1(b)}]$$

The above equations indicate that beam segment AX has a constant S.F. of magnitude $-R_A$; since the assumed S.F. system in Fig.5.6(b) forms a clockwise couple, the negative sign indicates that the S.F. in the segment is of negative nature. The B.M. indicated in the figure is sagging in nature; the positive sign of Eqn.I.1(b) shows that the assumed direction of moment at section X is correct. Thus, segment AX is subjected to a sagging moment. Further, the B.M. in segment AX varies linearly with x (Eqn.I.1(b)), with zero value at $A\,(x = 0)$, and a maximum value of $R_A\, a$ at $C\,(x = a)$. Substituting the value of $R_A\,(= P b/l)$, the maximum B.M. in segment $0 \leq x \leq a$ is given by

$$M_{\text{max}} = P\, a\, b/l$$

Figure 5.6(c) shows the F.B. diagram of the beam for section X at a distance $x > a$. The equilibrium conditions of segment AX yield

$$-R_A + P - F_x = 0 \quad \text{or} \quad F_x = P - R_A \qquad [I.2(a)]$$

and

$$R_A\,x - P\,(x-a) - M_x = 0 \qquad [\text{moments about } X]$$

or $\quad M_x = R_A\,x - P\,(x-a) \qquad$ [I.2(b)]

Since $P > R_A$, both F_x and M_x in Eqn.1.2 are positive; that is, the assumed directions of S.F. and B.M. are correct. Equation I.2(a) yields a constant value $(= R_B)$ for S.F. in the segment; the S.F. is positive as per the convention adopted.

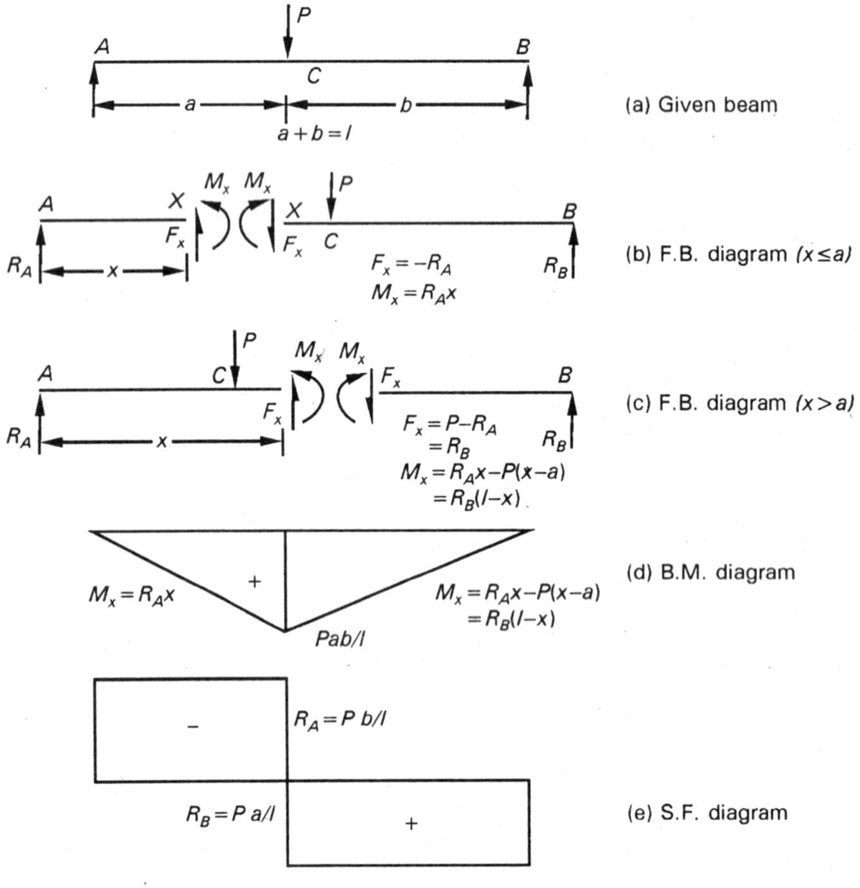

Fig.5.6 Simple beam under a concentrated load (Problem I)

The B.M. in the segment is positive (Eqn.I.2(b)) and varies linearly with x; the maximum value of M_x $(= P\,a\,b/l)$ occurs at $x = a$, and is zero for $x = l$. It may be noted that the B.M. is the same for both the segments at $x = a$, the loaded section. The B.M. diagram can be drawn to a suitable scale as indicated in Fig.5.6(d).

The S.F. at the loaded section is different for segments AC and CB. The S.F. has a constant value of $(-P\,b/l)$ for segment AC, and a constant value of $(+P\,a/l)$ for segment CB. There is a sudden change in the value of S.F. at C, the loaded section, as indicated in Fig.5.6(e). It should be noted that the sum of S.F. ordinates at C is P, the load applied at the section. Thus, S.F. at the loaded section has no meaning; S.F. is specified at an infinitesimal distance to the left or right of a loaded section (these values are always different).

The values of F and M can be developed by considering segment XB as well (Figs.5.6(b) and (c)); this will be particularly advantageous for section $x > a$. Students are advised to work out the problem again considering segment XB instead of segment AX.

Problem II Draw the B.M. and S.F. diagrams for a simple beam with a concentrated load at the mid-span section.

Solution II Because of the symmetry of loading on the beam, it is adequate to consider half of the beam and draw the diagrams symmetrically. In this case, we have

$$a = b = l/2 \quad \text{and} \quad R_A = R_B = P/2$$

Equations for B.M. and S.F. can be developed in the same way as explained in the previous problem. From Eqns.I.1(a) and I.2(a), we have a constant S.F. of $(-P/2)$ in segment AC, and $(+P/2)$ in segment CB. The maximum B.M. at mid-span section (the loaded section) can be obtained from Eqn.I.1(b) or I.2(b) as $(P\,l/4)$. The B.M. and S.F. diagrams are indicated in Figs.5.7(a) and (b), respectively.

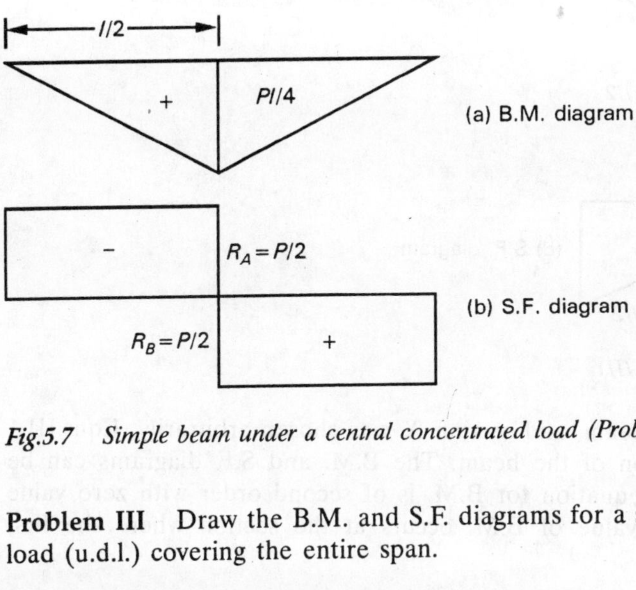

Fig.5.7 Simple beam under a central concentrated load (Problem II)

Problem III Draw the B.M. and S.F. diagrams for a simple beam with uniformly distributed load (u.d.l.) covering the entire span.

Solution III The beam and the loading are indicated in Fig.5.8(a). Because of the symmetry of the loading, we have $R_A = R_B = (wl/2)$. A section is taken at a distance x from support A; the F.B. diagrams of segments AX and XB are indicated in Fig.5.8(b). The equivalent loading on segment AX is also indicated in the figure. Considering segment AX, we have

$$-R_A + wx - F_x = 0 \qquad [R_A = wl/2]$$

or $\quad F_x = -w[(l/2) - x]$ [III.1]

Taking moments of the forces on segment AX about X, we obtain

$$R_A x - (wx)(x/2) - M_x = 0$$

Substituting for R_A in the above equation, we have

$$M_x = wx(l - x)/2$$ [III.2]

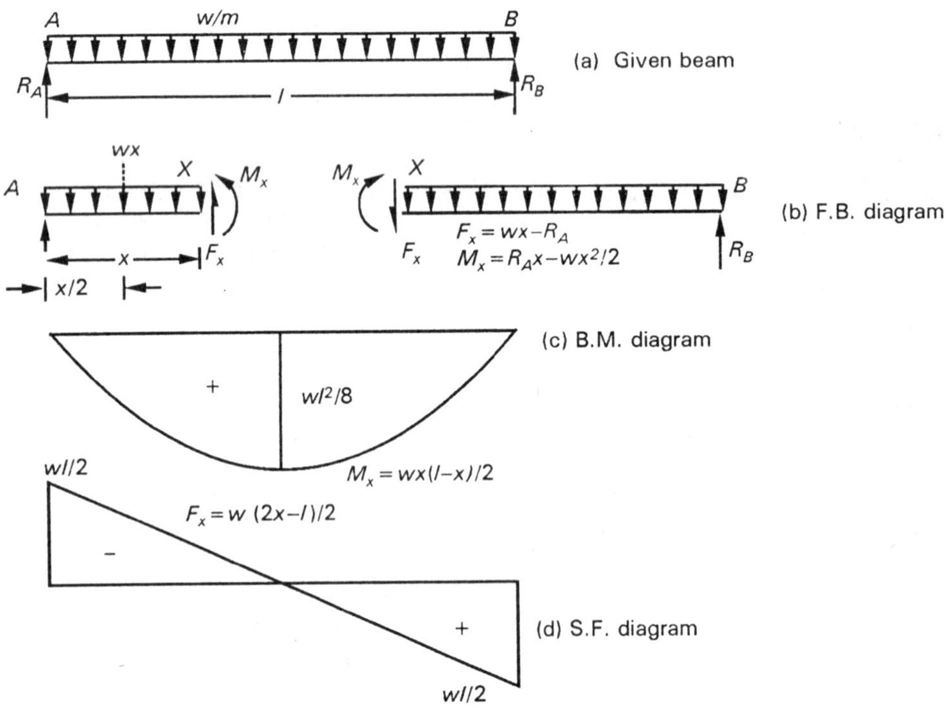

Fig.5.8 Simple beam under u.d.l. (Problem III)

Since the loading covers the entire beam, and section X was chosen arbitrarily, Eqns.III.1 and III.2 are applicable for any section of the beam. The B.M. and S.F. diagrams can be developed from these equations. The equation for B.M. is of second order with zero value at $x = 0$ and $x = l$. The maximum value of B.M. occurs at the section where the first differential of the expression for B.M. is zero; differentiating Eqn.III.2 and equating it to zero we obtain $x = l/2$. The maximum value of B.M. can be computed as $(wl^2/8)$. The B.M. diagram (parabolic profile) is indicated in Fig.5.8(c).

Equation III.1 is a linear function of x. The S.F. has a value of $(-wl/2)$ at $x = 0$, and the corresponding value at $x = l$ is $(+wl/2)$. The S.F. at the mid-span section $(x = l/2)$ is zero. The S.F. diagram is indicated in Fig.5.8(d).

Problem IV Draw the B.M. and S.F. diagrams for a simple beam with a load varying linearly along its span (Fig.5.9).

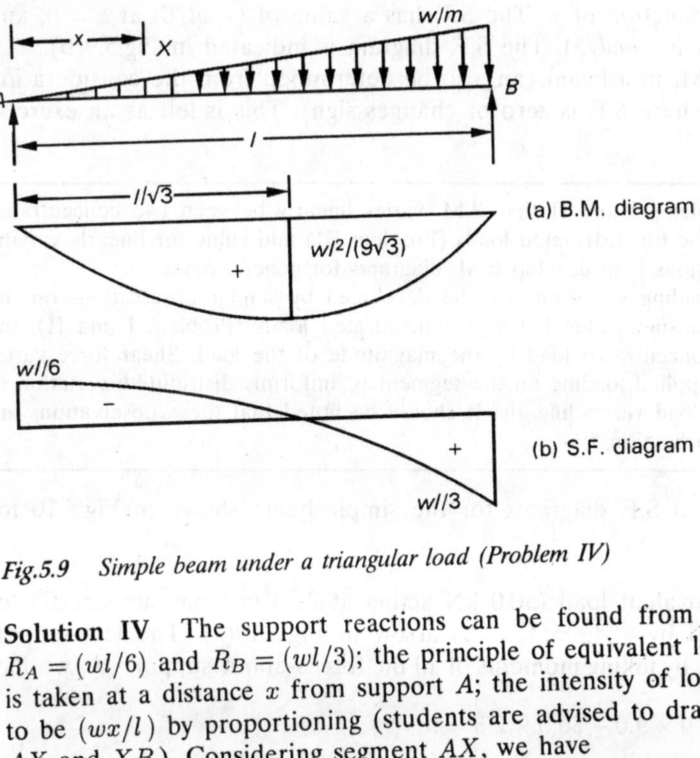

Fig.5.9 Simple beam under a triangular load (Problem IV)

Solution IV The support reactions can be found from static equilibrium conditions as $R_A = (wl/6)$ and $R_B = (wl/3)$; the principle of equivalent loading can be applied. A section is taken at a distance x from support A; the intensity of loading at section X can be found to be (wx/l) by proportioning (students are advised to draw the F.B. diagrams of segments AX and XB). Considering segment AX, we have

$$-R_A + w(x/l)(x/2) - F_x = 0$$

or $$F_x = -w(l^2 - 3x^2)/(6l) \quad \text{[IV.1]}$$

Taking moments of the forces on segment AX about X, we obtain

$$R_A x - (wx/l)(x/2)(x/3) - M_x = 0 \quad \text{or} \quad M_x = R_A x - wx^3/(6l)$$

Substituting for R_A in the above equation, we obtain

$$M_x = wx(l^2 - x^2)/(6l) \quad \text{[IV.2]}$$

The above equations are applicable for any section of the beam. The equation for B.M. is of third order with zero values at $x = 0$ and $x = l$. The maximum value of the B.M. occurs at a section where the first differential of Eqn.IV.2 has zero value; thus, we have

$$dM_x/dx = w(l^2 - 3x^2)/(6l) = 0 \quad \text{or} \quad x = l/\sqrt{3} = 0.577\,l$$

Substituting the above value of x in Eqn.IV.2, the maximum value of B.M. is given by

$$\text{Max. } M_x = 0.064\ 2\,w\,l^2$$

The corresponding B.M. diagram is indicated in Fig.5.9(a).

Equation IV.1 is a quadratic function of x. The S.F. has a value of $(-wl/6)$ at $x = 0$, and the corresponding value at $x = l$ is $(+wl/3)$. The S.F. diagram is indicated in Fig.5.9(b).

The location of maximum B.M. in a beam can also be determined from the consideration of S.F. (maximum B.M. occurs where S.F. is zero or changes sign). This is left as an exercise to students.

Note It can be noted from Problems I and II that B.M. varies linearly between two concentrated loads, while the variation is parabolic for distributed loads (Problem III) and cubic for linearly varying loads (Problem IV). These observations help develop B.M. diagrams for general cases.

The S.F. diagram for general loading on beams can be developed by similar observations on the previous problems. The S.F. has constant value between concentrated loads (Problem I and II); the value drops at the location of a concentrated load by the magnitude of the load. Shear force varies linearly over a segment when the applied loading on the segment is uniformly distributed; variation of S.F. is parabolic when the applied load varies linearly. It should be noted that these observations are based on the relations developed in Section 5.4.

Problem V Draw the B.M. and S.F. diagrams for the simple beam shown in Fig.5.10 for the given loading system.

Solution V The statically equivalent load (50.0 kN acting at 2.50 m from support B) for the distributed load is indicated by a discontinuous arrow in Fig.5.10(a). The left support reaction (R_A) can be determined by taking moments of all the forces about support B; we have

$$18.0R_A - 90.0 \times 14.0 - 150.0 \times 9.0 - 50.0 \times 2.5 = 0$$

or $\quad R_A = 151.94$ kN

$$R_B = (90.0 + 150.0 + 50.0) - RA = 138.06 \text{ kN}$$

In such cases comprising several concentrated loads, it is adequate to compute the B.M. at the locations of the loads. Referring to Fig.5.10(a), we have

$$M_C = R_A \times 4.0 = 151.94 \times 4.0 = 607.76 \text{ kNm}$$

$$M_D = 151.94 \times 9.0 - 90.0 \times 5.0 = 917.46 \text{ kNm}$$

$$M_E = 151.94 \times 13.0 - 90.0 \times 9.0 - 150.0 \times 4.0 = 565.22 \text{ kNm}$$

The variation of the B.M. between E and B will be parabolic because of the distributed loading over segment EB. It can be verified that the B.M. at section E is 565.22 kNm when computed for segment BE. The corresponding B.M. diagram is indicated in Fig.5.10(b).

From the definition, S.F. over segment AC should be equal to support reaction at A; the S.F. is negative because of the sign of the shear couple on adjacent faces of a section in segment AC. The computations for S.F. are indicated below.

S.F. over segment AC, $S_{AC} = -R_A = -151.9$ kN

S.F. to the right of C, $S_{CD} = S_{AC} + 90.0 = -61.94$ kN

S.F. to the right of D, $S_{DE} = S_{CD} + 150.0 = +88.06$ kN

S.F. to the left of B, $S_{BE} = R_B = +138.06$ kN

The S.F. diagram is indicated in Fig.5.10(c).

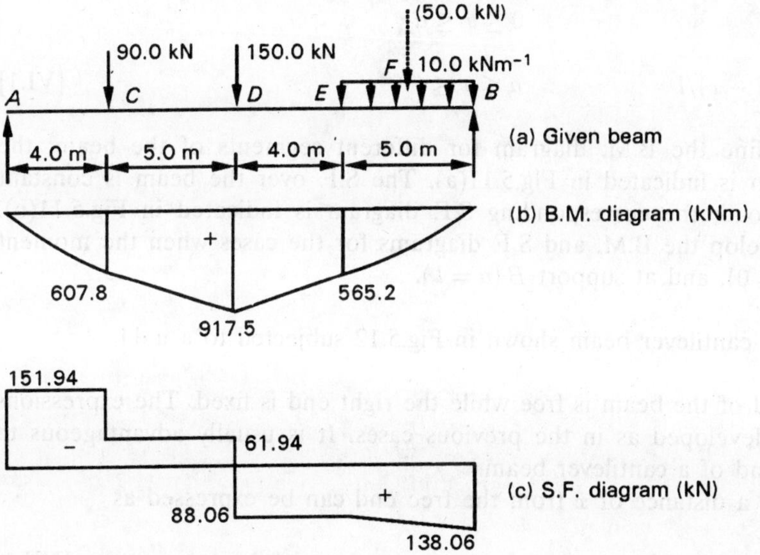

Fig.5.10 Simple beam under a given loading system (Problem V)

Problem VI Draw the B.M. and S.F. diagrams for a simple beam subjected to a moment at an intermediate section (Fig.5.11).

Solution VI Support reactions can be computed from simple statics as

$$R_A = -M/l \quad \text{and} \quad R_B = +M/l$$

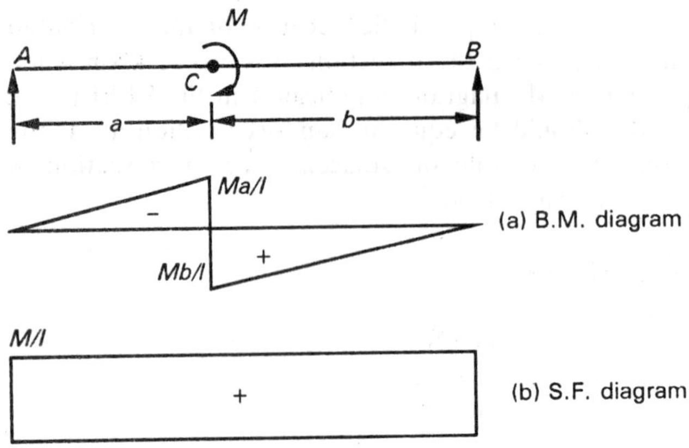

Fig.5.11 Simple beam with a moment (Problem VI)

The signs of the parameters pertain to the directions shown in Fig.5.11 (reaction at support A acts downwards, while that at B acts upwards). From the conventions adopted, we have

$$M_x = R_A\, x = -M\, x/l \qquad 0 \le x \le a$$

$$M_x = R_A\, x + M = M(l - x)/l \qquad a \le x \le l \qquad \text{[VI.1]}$$

The above equations define the B.M. diagram for different segments of the beam; the corresponding B.M. diagram is indicated in Fig.5.11(a). The S.F. over the beam is constant since there is no applied load; the corresponding S.F. diagram is indicated in Fig.5.11(b). Students are advised to develop the B.M. and S.F. diagrams for the cases when the moment is applied at support $A\,(a = 0)$, and at support $B\,(a = l\,)$.

Problem VII Analyse the cantilever beam shown in Fig.5.12 subjected to a u.d.l.

Solution VII The left end of the beam is free while the right end is fixed. The expressions for B.M. and S.F. can be developed as in the previous cases. It is usually advantageous to assume origin at the free end of a cantilever beam.

The B.M. at a section at a distance of x from the free end can be expressed as

$$M_x = -w\, x\,(x/2) = -w\, x^2/2 \qquad \text{[VII.1]}$$

The above equation indicates that the B.M. varies parabolically from zero at the free end $(x = 0)$ to an extreme value of $(-w\, l^2/2)$ at the fixed end $(x = l\,)$. The B.M. diagram is indicated in Fig.5.12(a).

The S.F. at the same section can be expressed as

$$F_x = w\, x \qquad \text{[VII.2]}$$

It can be noted from the above equation that the S.F. varies linearly from zero at the free end to a maximum value of $w\, l$ at the fixed end. The S.F. diagram is indicated in Fig.5.12(b).

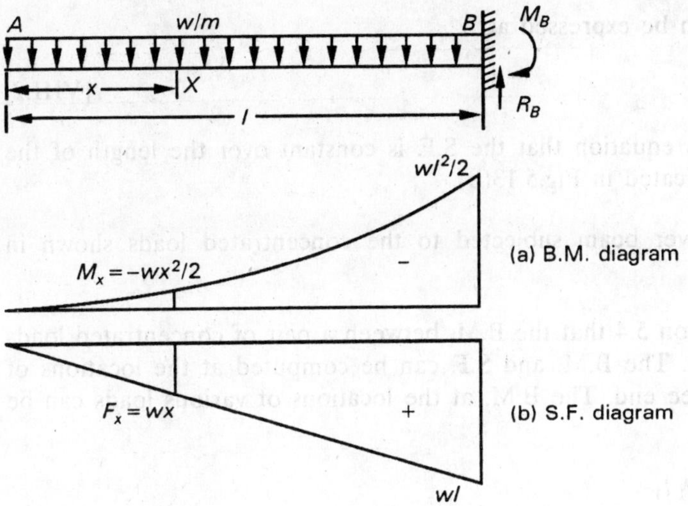

Fig.5.12 Cantilever with u.d.l. (Problem VII)

Problem VIII Analyse the cantilever beam shown in Fig.5.13 subjected to a concentrated force at the free end.

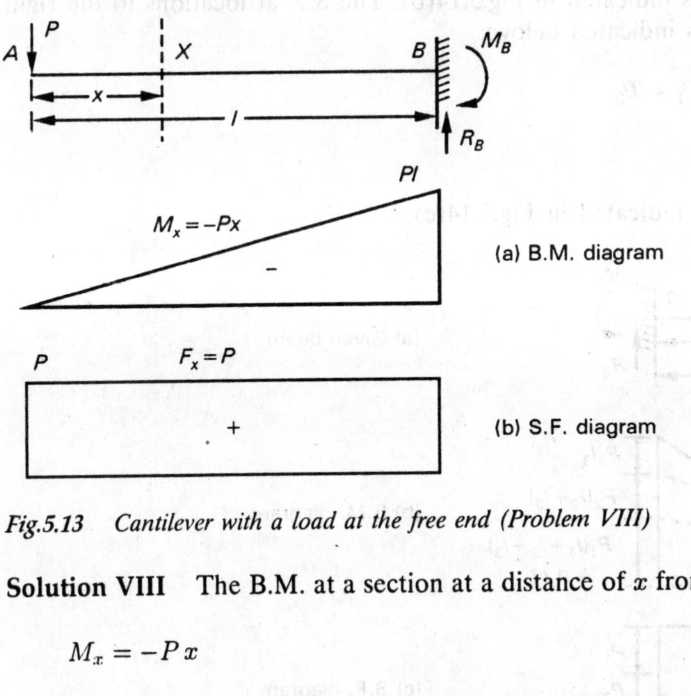

Fig.5.13 Cantilever with a load at the free end (Problem VIII)

Solution VIII The B.M. at a section at a distance of x from the free end can be expressed as

$$M_x = -Px \quad \text{[VIII.1]}$$

The above equation indicates that the B.M. varies linearly from zero at the free end ($x = 0$) to an extreme value ($-Pl$) at the fixed end ($x = l$). The B.M. diagram is indicated in Fig.5.13(a).

The S.F. at the same section can be expressed as

$$F_x = P \qquad \text{[VIII.2]}$$

It can be noted from the above equation that the S.F. is constant over the length of the cantilever. The S.F. diagram is indicated in Fig.5.13(b).

Problem IX Analyse the cantilever beam subjected to the concentrated loads shown in Fig.5.14.

Solution IX We know from Section 5.4 that the B.M. between a pair of concentrated loads is linear, while the S.F. is constant. The B.M. and S.F. can be computed at the locations of the loads commencing from the free end. The B.M. at the locations of various loads can be computed as indicated below:

$$M_A = 0.0 \qquad M_C = -P_1\, l_1$$

$$M_D = -P_1\,(l_1 + l_2) - P_2\, l_2$$

$$M_B = -P_1\,(l_1 + l_2 + l_3) - P_2\,(l_2 + l_3) - P_3\, l_3$$

The corresponding B.M. diagram is indicated in Fig.5.14(b). The S.F. at locations to the right of given loads can be computed as indicated below:

$$F_A = +P_1 \qquad F_C = +P_1 + P_2$$

$$F_D = +P_1 + P_2 + P_3 = F_B$$

The corresponding S.F. diagram is indicated in Fig.5.14(c).

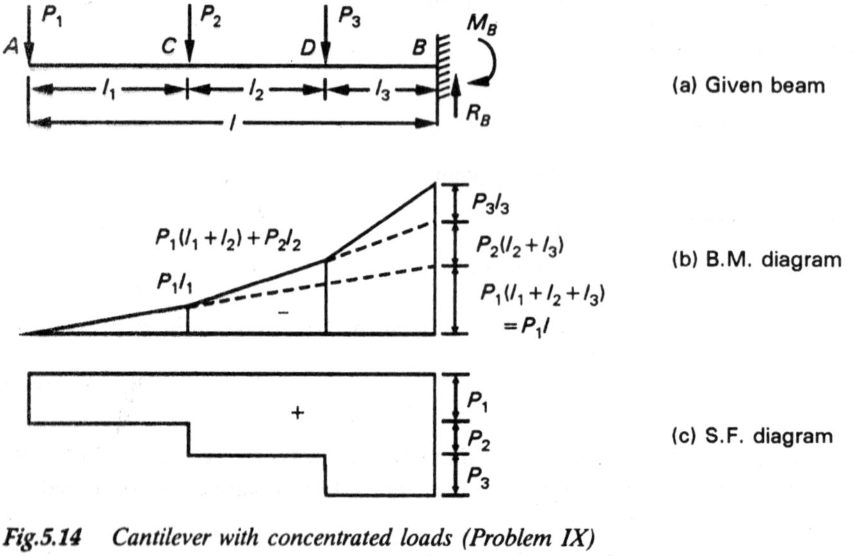

Fig.5.14 Cantilever with concentrated loads (Problem IX)

Problem X Analyse the cantilever shown in Fig.5.15 with the load varying linearly along its span.

Solution X It is convenient to assume origin at the free end of the cantilever. Taking a section at a distance x from the free end of the cantilever, the intensity of loading at section X (distance x from the origin) can be found to be $(w x/l)$. Considering segment BX, we have

$$F_x = -w x^2/(2 l) \quad \text{[X.1]}$$

$$M_x = -w x^3/(6 l) \quad \text{[X.2]}$$

Equation X.2 for B.M. is of third order with zero value at $x = 0$ and a maximum value of $(-w l^2/6)$ at $x = l$. The corresponding B.M. diagram is indicated in Fig.5.15(a). The S.F. is a quadratic function of x (Eqn.X.1). It has a zero value at $x = 0$, and a value of $(-wl/2)$ at $x = l$. The corresponding S.F. diagram is indicated in Fig.5.15(b).

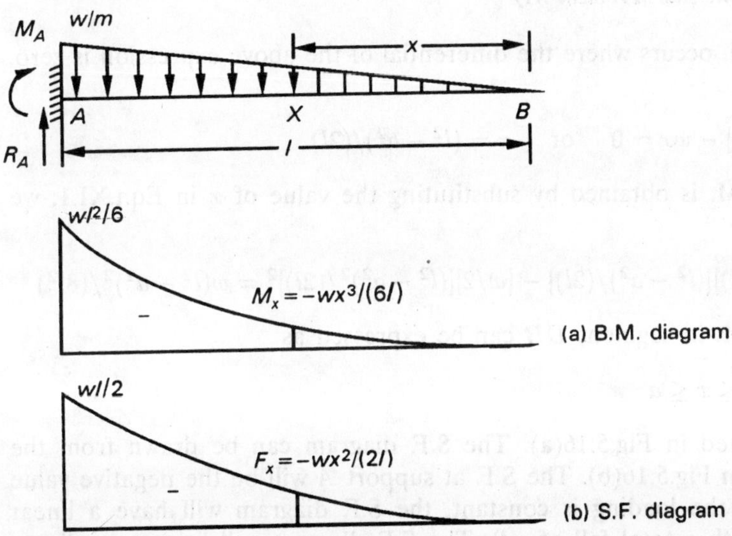

Fig.5.15 Cantilever under a triangular loading (Problem X)

Problem XI Analyse the beam with an overhang shown in Fig.5.16.

Solution XI The support reactions can be computed from statics as

$$R_A = w(l^2 - a^2)/(2l) \quad \text{and} \quad R_B = w(l + a)^2/(2l)$$

Taking the origin at support A, the B.M. in segment AB can be expressed as

$$M_x = [w(l^2 - a^2)/(2l)]x - wx^2/2 \qquad 0 \leq x \leq 1 \quad \text{[XI.1]}$$

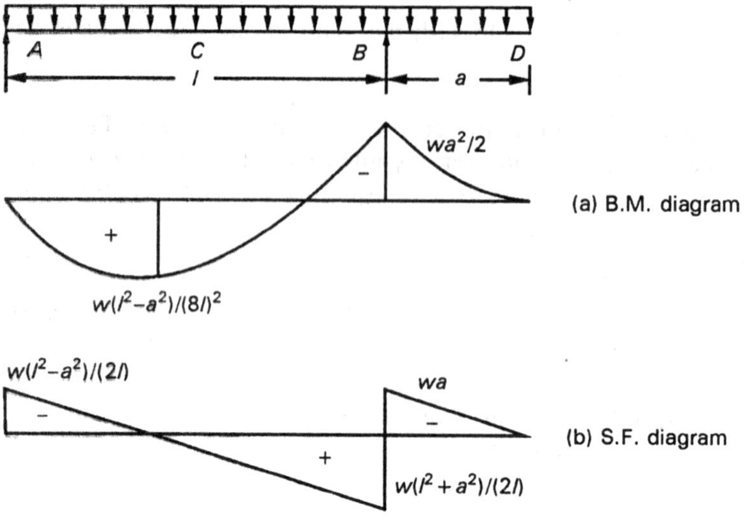

Fig.5.16 *Beam with overhang on one side (Problem XI)*

The maximum value of the B.M. occurs where the differential of the above expression is zero. Thus, we have

$$dM_x/dx = [w(l^2 - a^2)/(2l)] - wx = 0 \quad \text{or} \quad x = (l^2 - a^2)/(2l)$$

The maximum value of the B.M. is obtained by substituting the value of x in Eqn.XI.1; we have, accordingly

$$\text{Max. } M_x = [w(l^2 - a^2)/(2l)][(l^2 - a^2)/(2l)] - [w/2][(l^2 - a^2)^2/(2l)]^2 = w(l^2 - a^2)^2/(8l^2)$$

Taking the origin at D, the B.M. in segment DB can be expressed as

$$M_x = -wx^2/2 \qquad 0 \le x \le a$$

The B.M. diagram is indicated in Fig.5.16(a). The S.F. diagram can be drawn from the support reactions as indicated in Fig.5.16(b). The S.F. at support A will be the negative value of the support reaction. Since the loading is constant, the S.F. diagram will have a linear variation between A and B (with a total fall of wl). The S.F. diagram will have an ordinate $R_B = [w(l+a)^2/(2l)]$ at B. The S.F. diagram has a constant slope again over segment BD, with a zero value at D.

5.5 Summary

Beams are subjected to loads transverse to their axes, and develop B.M. and S.F. due to the loads applied. The concepts of B.M. and S.F. are explained in this chapter. A beam is statically determinate when it can be analysed using the equations of static equilibrium alone; otherwise it is termed indeterminate (or redundant). The procedure of computing B.M. and S.F. at various sections, and their graphical representation is discussed in this chapter. Various types of beams subjected to typical loading cases are considered in the analysis.

5.6 Exercises

5.1 Discuss the concepts of B.M. and S.F.

5.2 Develop the relation between B.M. and S.F. in a beam.

5.3 Indicate the relation between S.F. and loading function of a beam.

5.4 Explain how the extreme values of B.M. in a beam can be determined.

5.5 Develop B.M. and S.F. diagrams for the beams shown in Figs.E 5.1 indicating the maximum and minimum values.

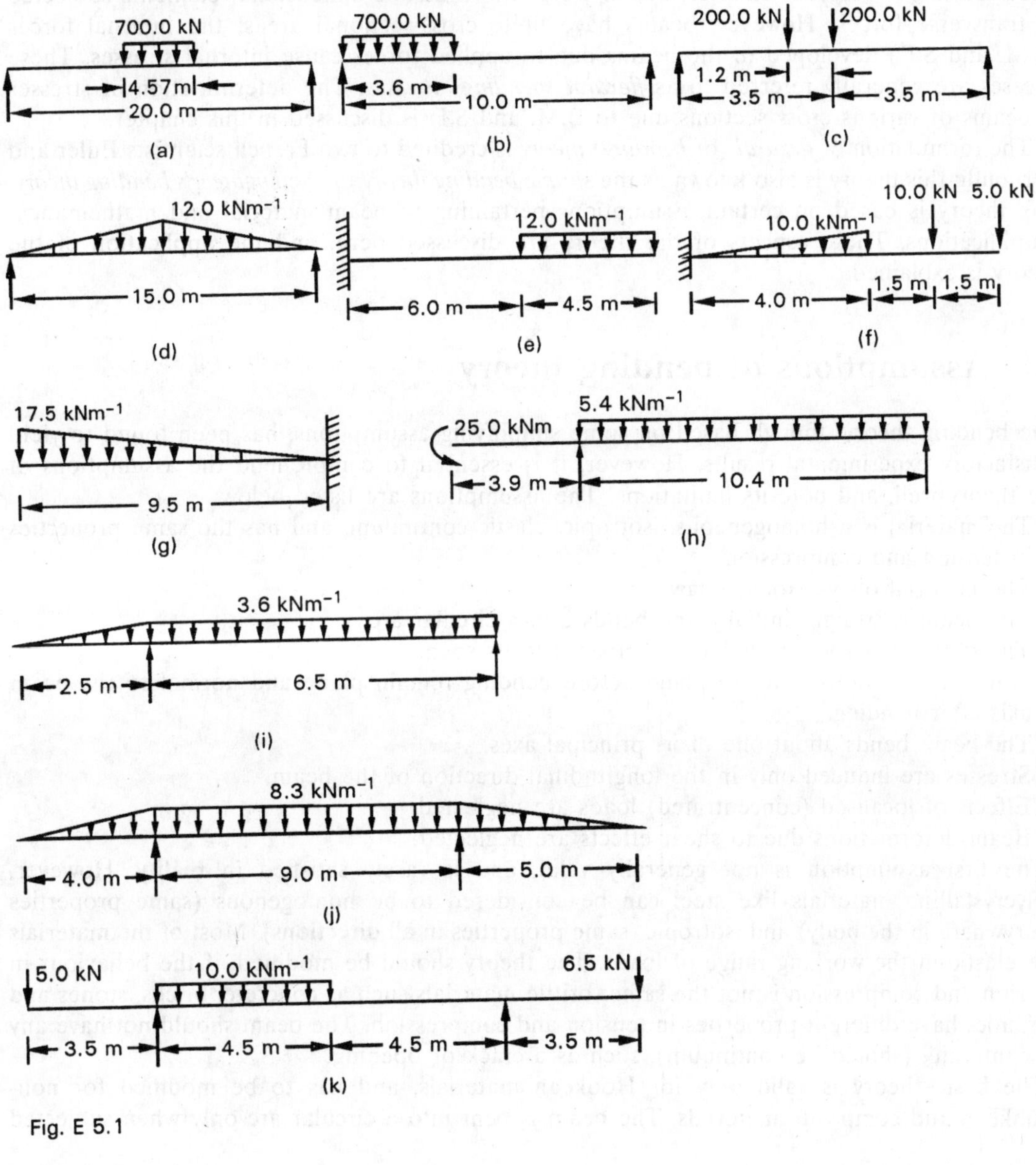

Fig. E 5.1

6 Flexural analysis of beams

6.0 Introduction

The analysis of beams involving the determination of B.M. and S.F. at various sections was discussed in the previous chapter; beams were shown as one-dimensional elements subjected to transverse forces. However, beams have finite cross-sectional areas; the internal forces (B.M. and S.F.) developed in the beams due to applied forces cause internal stresses. These stresses are generally referred to as *flexural (bending) stresses*. The determination of stresses in beams of various cross sections due to B.M. and S.F. is discussed in this chapter.

The formulation of *flexural* (or *bending*) *theory* is credited to two French scientists Euler and Bernoulli; this theory is also known as the *simple bending theory* or the *engineer's bending theory*. The theory is based on certain assumptions pertaining to beam material and mathematical simplifications. These aspects of the theory are discussed here, and the application of the theory is explained.

6.1 Assumptions of bending theory

The bending theory, though based on some simplifying assumptions, has been found to yield satisfactory experimental results. However, it is essential to comprehend the assumptions of the theory well, and note its limitations. The assumptions are listed below:

1. The material is a homogeneous, isotropic, elastic continuum, and has the same properties in tension and compression.
2. The material obeys Hooke's law.
3. The beam is straight initially and bends into a circular arc.
4. The radius of curvature is large compared to its span.
5. Transverse sections that are plane before bending remain plane and normal to the beam axis after bending.
6. The beam bends about one of its principal axes.
7. Stresses are induced only in the longitudinal direction of the beam.
8. Effects of localised (concentrated) loads are neglected.
9. Beam deformations due to shear effects are neglected.

The first assumption is not generally valid, and is rarely satisfied in reality. However, polycrystalline materials like steel can be considered to be homogenous (same properties everywhere in the body) and isotropic (same properties in all directions). Most of the materials are elastic in the working range of loads. The theory should be modified, if the behaviour in tension and compression is not the same; brittle materials such as concrete, bricks, stones and ceramics have different properties in tension and compression. The beam should not have any discontinuity (should be continuum), such as a crack or opening.

The basic theory is valid only for Hookean materials, and has to be modified for non-Hookean and composite materials. The beam is bent into a circular arc only when subjected to constant B.M. A large radius of curvature (compared to beam span) implies small deflections (normal to the beam axis). However, the deflections are small so that the geometry of the structure is not altered. The assumption that sections which are plane before bending remaining plane after bending indicates that the section is not distorted due to bending. In other words, strains vary linearly along the beam depth.

The beam is assumed to be symmetrical about the axis of loading in its cross section (bending about a principal axis). Further, the stresses in vertical direction due to applied loads are assumed to be negligible (assumption 7). Effects of localised concentrated forces (applied loads and support reactions) are not considered by this theory. The deformation of the beam due to S.F. is not considered by Euler–Bernoulli's theory.

6.2 Euler–Bernoulli's bending equation

Consider a beam element shown in Fig.6.1(a). The element is bent into a circular arc of angle $d\theta$ with O' as the centre. The beam may be of an arbitrary cross section shown in Fig.6.1(b); the section is symmetric about the Y-axis. The upper layers of the beam contract and the lower layers elongate due to bending. The strains are distributed linearly along the depth of the beam as per the fifth assumption (Fig.6.1(c)); tensile strains are positive and compressive strains are negative. The strain distribution indicates that there should be a layer along which strains are zero. This layer is called the neutral axis of the beam, and is shown as nn in Fig.6.1(a). The radius of the arc of bending is shown as R upto the *neutral axis*. Considering layer aa at a distance y from the neutral axis (Fig.6.1(a)), we have

$$nn = R\,d\theta, \qquad aa = (R+y)\,d\theta$$

The strain in layer aa due to bending can be expressed as

$$\varepsilon = (aa - nn)/nn = [(R+y)\,d\theta - R\,d\theta]/(R\,d\theta) = y/R$$

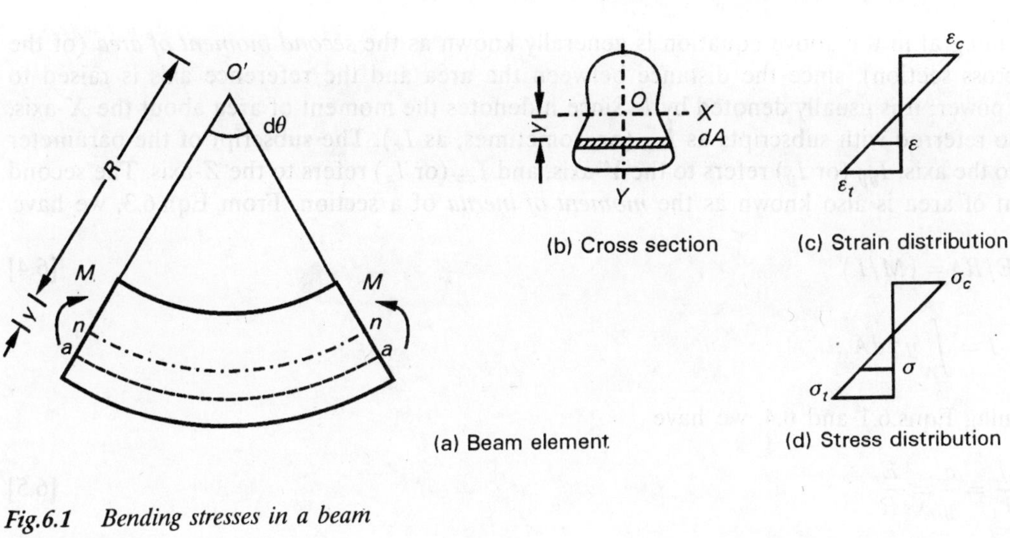

(b) Cross section

(c) Strain distribution

(d) Stress distribution

(a) Beam element

Fig.6.1 Bending stresses in a beam

Expressing strain in terms of stress (Hooke's !aw), we have

$$[\sigma/E] = [y/R]$$

or $$[\sigma/y] = [E/R] \qquad [6.1]$$

Since the section is not subjected to any direct force, the total force on the section should be zero (force equilibrium condition). Considering an infinitesimal area dA at a distance y from the neutral axis, and applying the force equilibrium condition, we have

$$\int_A \sigma \, dA = 0 \qquad \text{[integral over the sectional area]}$$

Substituting for σ from Eqn.6.1 in the above equation, we have

$$(E/R)\int_A y \, dA = 0 \quad \text{or} \quad \int_A y \, dA = 0 \qquad [6.2]$$

The above equation indicates that the neutral axis passes through the centroid of the beam section. Further, the moment of forces on the beam section should be equal to M, the applied moment (moment equilibrium condition). Taking moment of the forces on the section about its neutral axis, we have

$$\int_A \sigma y \, dA = M$$

Substituting for σ from Eqn.6.1 in the above equation, we have

$$(E/R)\int_A y^2 \, dA = M \qquad [6.3]$$

The integral in the above equation is generally known as the *second moment of area* (of the beam cross section), since the distance between the area and the reference axis is raised to second power; it is usually denoted by I. Since it denotes the moment of area about the X-axis, it is also referred with subscripts as I_{xx} (or, sometimes, as I_x). The subscript of the parameter refers to the axis; I_{yy} (or I_y) refers to the Y-axis, and I_{zz} (or I_z) refers to the Z-axis. The second moment of area is also known as the *moment of inertia* of a section. From Eqn.6.3, we have

$$(E/R) = (M/I) \qquad [6.4]$$

where $I = \int_A y^2 \, dA$

Combining Eqns.6.1 and 6.4, we have

$$\frac{M}{I} = \frac{\sigma}{y} = \frac{E}{R} \qquad [6.5]$$

The above equation is known as *Euler–Bernoulli's bending equation* (or simply, *bending equation*). It is essential to note that the moment of forces on the section about the Y-axis is zero (since no moment acts normal to the plane of bending). Thus, we have

$$\int_A \sigma\, x\, dA = 0$$

Substituting for σ from Eqn.6.1 in the above equation, we have

$$(E/R)\int_A x\, y\, dA = 0$$

or $$\int_A x\, y\, dA = 0 \qquad [6.6]$$

The integral in the above equation is generally known as the *product of inertia* (of the beam cross section), and is denoted by I_{xy}. A zero value of the parameter indicates symmetry of the section about at least one of the axes (Y-axis, in this case). If the load is applied on the beam in any direction other than the Y-axis, the load should be resolved into the X- and Y-components before applying Eqn.6.5 about the respective axes.

6.3 Moment of inertia

The parameter, moment of inertia (M.I.), plays a significant role in structural analysis. The behaviour of a member subjected to B.M. is governed by its M.I. The determination of M.I. is simplified by noting the two associated theorems.

6.3.1 Theorem of parallel axes

Figure 6.2 refers to an arbitrary cross section of area A with the X-axis passing through its centroid. Considering an infinitesimal element dA at a distance y from the X-axis, we have

$$I_x = \int y^2 dA \qquad \text{[Eqn.6.4]}$$

From the definition, the M.I. about an axis BB at a distance b from the X-axis is given by

$$\begin{aligned} I_b &= \int (y+b)^2\, dA \\ &= \int y^2\, dA + b^2 \int dA + 2b \int y\, dA \end{aligned}$$

However, the third component is zero, being the moment of area about the centroid of the cross section (Eqn.6.2). Thus, we have

$$I_b = I_x + A\, b^2 \qquad [6.7]$$

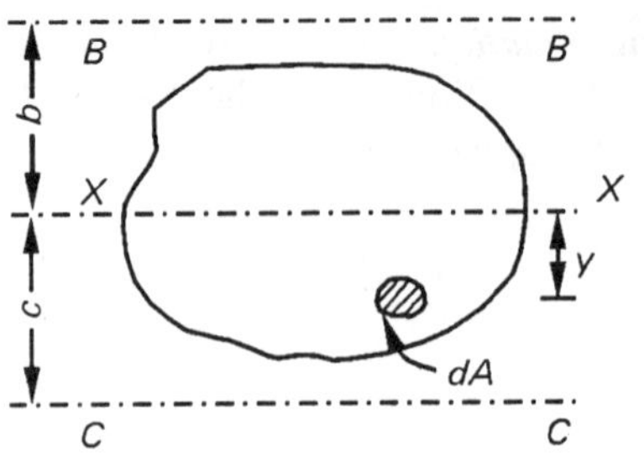

Fig.6.2 Parallel axes theorem

That is, if the M.I. of a section about its centroidal axis is known, the M.I. about any axis parallel to the centroidal axis is given by the sum of the M.I. about the centroidal axis and the product of the area and the square of the distance between the parallel axes. This is known as the *theorem of parallel axes*. The M.I. about axis CC at a distance c from the X-axis (Fig.6.2) is given by

$$I_c = I_x + A\,c^2$$

Combining with Eqn.6.7, we obtain

$$I_c = [I_b - A\,b^2] + A\,c^2 \qquad [6.8]$$

The above equation indicates that the M.I. about an axis should be first transferred to the centroidal axis before computing the value about another parallel axis.

6.3.2 Theorem of perpendicular axes

Figure 6.3 depicts three coordinate axes through the centroid of an arbitrary section. The section lies in the X–Y plane, while the Z-axis is normal to the section. Considering an elemental area dA at a distance r from the origin, the M.I. of the section about the Z-axis can be computed as

$$I_z = \int r^2\,dA = \int (x^2 + y^2)\,dA$$

where x and y are the coordinates of the elemental area.

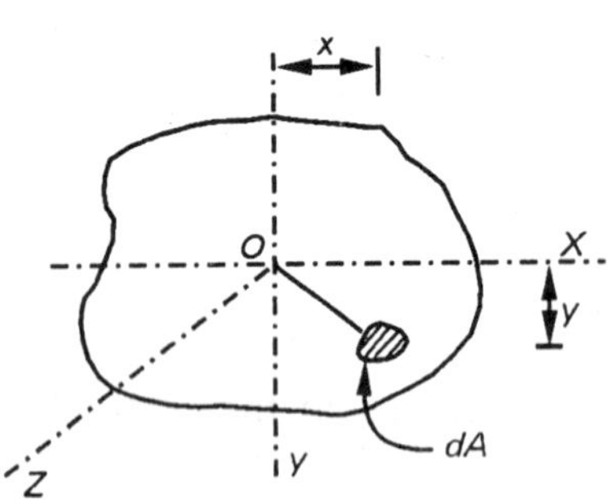

Fig.6.3 Perpendicular axes theorem

The above equation can be expressed as

$$I_z = I_x + I_y \qquad [6.9]$$

where I_x = M.I. about the X-axis and I_y = M.I. about the Y-axis.

In other words, the M.I. about an axis is the sum of the M.I. about the two other perpendicular axes. The M.I. of a section can be computed from these theorems. The moments of area of a few typical sections are computed for illustration, along with their application in determining flexural stress distribution.

Problem I Determine the M.I. of the sections shown in Fig.6.4 about neutral axis.

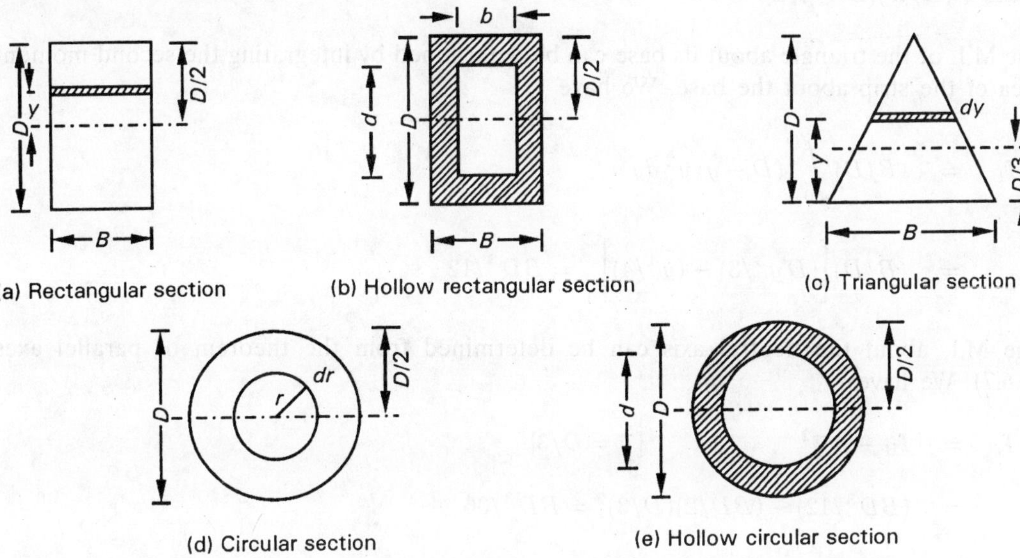

Fig.6.4 Moments of inertia of a few typical cross sections (Problem I)

Solution I ***(a) Rectangular section (Fig.6.4(a))*** The section is of width B and depth D; the centroid of the section is at mid-depth due to symmetry as indicated in Fig.6.4(a). Considering a strip of thickness dy at a distance y from the centroid, and noting that y varies from $-D/2$ to $+D/2$, we have

$$I_x = \int_{-D/2}^{D/2} y^2 B\, dy \qquad [dA = B\, dy]$$

$$= B\left[\frac{y^3}{3}\right]_{-D/2}^{D/2} = BD^3/12$$

The M.I. about the Y-axis can be determined as

$$I_y = B^3D/12 \qquad \left[I_y = \int_A x^2 dA\right]$$

(b) Hollow rectangular section (Fig.6.4(b)) The hollow rectangular section shown in Fig.6.4(b) has an external width B and depth D, the corresponding internal dimensions being b and d. The net M.I. about the X-axis through its centroid can be determined by subtracting the M.I. of the internal rectangle (hollow part) from that of the external rectangle. Thus, we have

$$I_x = (B\,D^3 - b\,d^3)/12$$

(c) Triangular section (Fig.6.4(c)) The section has a base width B and height D as indicated in Fig.6.4(c). Considering a strip at a distance y from the base, the area of the strip can be expressed as

$$dA = (B/D)(D - y)\,dy$$

The M.I. of the triangle about its base can be determined by integrating the second moment of area of the strip about the base. We have

$$\begin{aligned} I_B &= (B/D)\int_0^D (D - y)\,y^2\,dy \\ &= (B/D)\Big[(D\,y^3/3) - (y^4/4)\Big]_0^D = BD^3/12 \end{aligned}$$

The M.I. about the neutral axis can be determined from the theorem of parallel axes (Eqn.6.7). We have

$$\begin{aligned} I_x &= I_B - A\overline{y}^2 \qquad [\overline{y} = D/3] \\ &= (BD^3/12) - (BD/2)(D/3)^2 = BD^3/36 \end{aligned}$$

(d) Circular section (Fig.6.4(d)) Considering a ring of radius r and thickness dr in the circular section of diameter D indicated in Fig.6.4(d), the M.I. of the ring about the axis normal to the plane of the figure through the centre of the circle can be expressed as

$$dI_z = 2\pi \int_0^R r\,dr(r^2)$$

The M.I. of the section can be obtained by integrating the above equation in the limits 0 to $R(= D/2)$. We obtain

$$I_z = 2\pi \int_0^R r^3\,dr = \pi\,R^4/2 = \pi\,D^4/32$$

The M.I. of the section about the X- and Y-axes is equal because of the symmetry of the section. Applying the theorem of perpendicular axes, we have

$$I_z = I_x + I_y = 2I_x \quad \text{or} \quad I_x = I_y = I_z/2 = \pi\,D^4/64 = \pi\,R^4/4$$

(e) Hollow circular section (Fig.6.4(e)) The M.I. can be obtained the same way as explained for a hollow rectangular section (Fig.6.4(b)). We have

$$I_z = \pi(D^4 - d^4)/32, \qquad I_x = I_y = \pi(D^4 - d^4)/64$$

Problem II Determine the M.I. of the *I*-section shown in Fig.6.5.

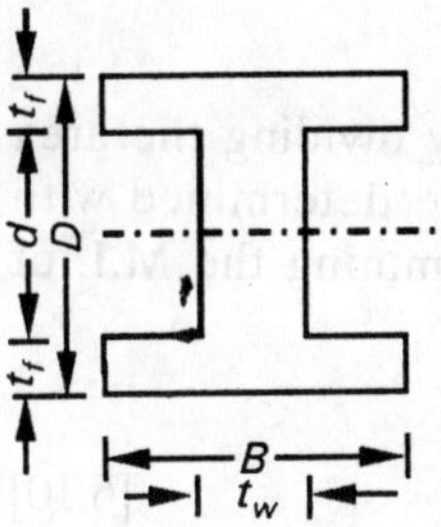

Fig.6.5 Symmetric I-section (Problem II)

Solution II The section is symmetric about the X- and Y-axes. The neutral axis lies at mid-depth for the computations of M.I. about the X-axis as indicated in Fig.6.5. The section may be visualised as a rectangle of size $B \times D$ with another rectangle of size $(B - t_w) \times d$ removed. The M.I. can be computed accordingly as

$$I_x = [BD^3/12] - [(B - t_w)\, d^3/12]$$

Problem III Determine the M.I. of a circular ring of diameter D and thickness t.

Solution III The M.I. of the ring can be computed from the expression developed for the annular ring of Fig.6.4(e) (Problem I(e)). The circular ring shown in Fig.6.6 has a mean diameter of D and a thickness of t. The M.I. of an annular ring of external diameter D and internal diameter d can be simplified as indicated below.

$$\begin{aligned} I_x &= \pi(D^4 - d^4)/64 = \pi(D^2 + d^2)(D + d)(D - d)/64 \\ &= \pi D^3 t/8 \qquad \text{[since } D \simeq d \text{ and } (D - d) \simeq 2t] \end{aligned}$$

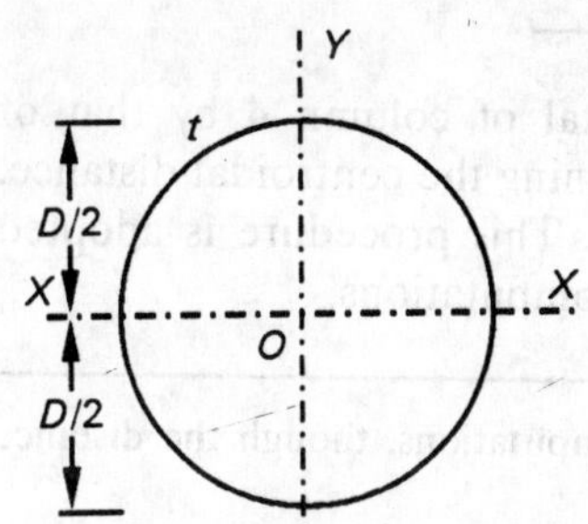

Fig.6.6 Moment of inertia of a circular ring (Problem III)

The above expression can also be developed by considering the M.I. of the ring about the Z-axis as in Problem I(d). We have

$$I_x = \pi D t (D/2)^2/2 = \pi D^3 t/8$$

6.4 Moments of inertia of arbitrary sections

The moment of inertia (M.I.) of an arbitrary section can be determined by dividing the area into smaller regular segments. The area and centroid of the section can be determined with the help of these segments. The M.I. of the section can be obtained by summing the M.I. of these segments about the required axis. Mathematically, we have

$$I = \sum_{i=1}^{n} [I_i' + (a_i \, y_i'^2)] \quad [6.10]$$

where I' = M.I. about the segment's centroidal axis

a = area of the segment

y' = distance between the centroidal axes of the segment and the section

n = the number of segments, and

i = subscript referring to the segment

The M.I. of a segment can be determined using the expressions developed so far. The computations can be tabulated as indicated below in a compact form.

Segment No. (1)	a (m^2) (2)	y (m) (3)	ay (m^3) (4)	$a(y-\overline{y})^2$ (m^4) (5)	I' (m^4) (6)
Total					

The location of centroid ($\overline{y}$) is obtained by dividing the total of column 4 by that of column 2; computations of column 5 can be carried out after obtaining the centroidal distance. The M.I. is obtained by adding up the total of columns 5 and 6. This procedure is adopted in determining the M.I. of the irregular sections in subsequent computations.

Note The distance y in column 3 refers to top fibre in subsequent computations, though the distance can be measured from top or bottom fibre of the section.

Problem IV Determine the M.I. of the unsymmetrical I-section shown in Fig.6.7.

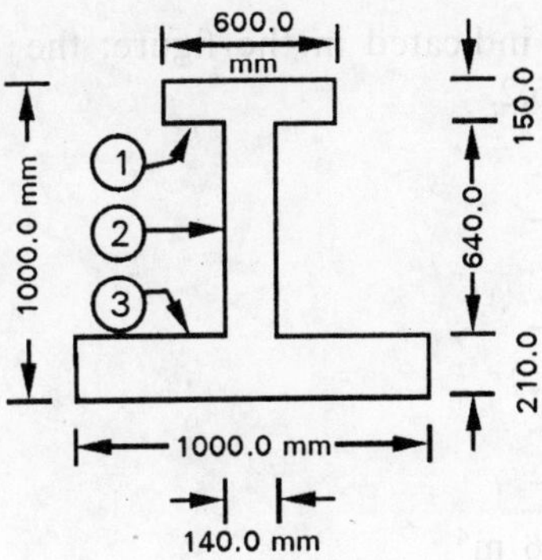

Fig.6.7 Unsymmetric I-section (Problem IV)

Solution IV The section can be divided into three rectangular segments, namely the top flange, web and the bottom flange. The computations are shown in Table 6.1.

Table 6.1 *Computations for M.I. of an I-section (Fig.6.7)*

Segment No. (1)	a (m^2) (2)	y (m) (3)	ay (m^3) (4)	$a(y-\overline{y})^2$ (m^4) (5)	I' (m^4) (6)
1	0.60 × 0.15 = 0.090	0.075	0.007	0.025 6	0.000 2
2	0.14 × 0.64 = 0.090	0.470	0.042	0.001 7	0.003 1
3	1.00 × 0.21 = 0.210	0.895	0.188	0.017 3	0.000 8
Total	0.390		0.237	0.044 6	0.004 1

$\overline{y} = 0.237/0.390 = 0.608$ m

$I = 0.044\,6 + 0.004\,1 = 0.048\,7$ m^4

Problem V Determine the maximum stresses in a beam of span 12.0 m supporting a distributed load of 60.0 kNm^{-1}. The cross section of the beam is shown in Fig.6.8; the material has a density of 25.0 kNm^{-3}.

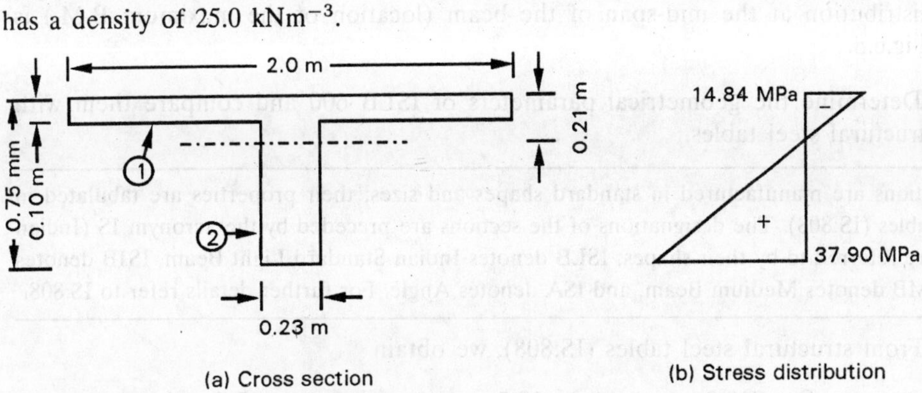

Fig.6.8 Stresses in a T-beam (Problem V)

Solution V Geometrical properties of the cross section, and the maximum B.M. in the beam are computed first before applying the Euler–Bernoulli bending theory.

The cross section is divided into two rectangular segments as indicated in the figure; the computations for the geometrical properties are shown in Table 6.2.

Table 6.2 *Computations for M.I. of a T-section (Fig.6.8)*

Segment No.	a (m^2)	y (m)	ay (m^3)	$a(y-\overline{y})^2$ (m^4)	I' (m^4)
1	2.00 × 0.10= 0.200	0.050	0.010	0.005 2	0.000 2
2	0.23 × 0.65= 0.150	0.425	0.064	0.006 9	0.005 3
Total	0.350		0.074	0.012 1	0.005 5

$\overline{y} = 0.074/0.350 = 0.211$ m; $I = 0.012\ 1 + 0.005\ 5 = 0.017\ 6\ m^4$

$$
\begin{aligned}
\text{Self-weight of the beam} &= A \times \rho \text{ per metre} = 0.35 \times 25.0 = 8.75 \text{ kNm}^{-1}\\
\text{Total load on the beam} &= \text{Imposed load} + \text{self-weight}\\
&= (60.0 + 8.75) = 68.75 \text{ kNm}^{-1}\\
\text{Maximum B.M. in the beam} &= w\,l^2/8 = 68.75 \times 12.0^2/8 \text{ kNm}\\
&= 1\ 237.5 \text{ kNm} = 1.237\ 5 \text{ MNm}\\
\text{Stress in the beam}\quad \sigma &= M_{\max}\,y/I = [1.237\ 5\,y/0.017\ 6] \text{ MPa} \qquad \text{[Eqn.6.5]}\\
&= 70.31\,y \text{ MPa} \qquad (y \text{ in metres})
\end{aligned}
$$

Stresses are computed at the top fibre ($y = -0.211$ m from the neutral axis), and at the bottom fibre ($y = +0.539$ m).

$$
\begin{aligned}
\text{Stress at top fibre} &= 70.31 \times (-0.211) \text{ MPa} = -14.84 \text{ MPa} \quad \text{(compressive)}\\
\text{Stress at bottom fibre} &= 70.31 \times (0.539) \text{ MPa} = +37.90 \text{ MPa} \quad \text{(tensile)}
\end{aligned}
$$

The stress distribution at the mid-span of the beam (location of the maximum B.M.) is also shown in Fig.6.8.

Problem VI Determine the geometrical parameters of ISLB 600 and compare them with the values of structural steel tables.

Note Steel sections are manufactured in standard shapes and sizes; their properties are tabulated in structural steel tables (IS:808). The designations of the sections are preceded by the acronym IS (Indian Standard), and are referred to by their shapes; ISLB denotes Indian Standard Light Beam, ISJB denotes Junior Beam, ISMB denotes Medium Beam, and ISA denotes Angle. For further details refer to IS:808.

Solution VI From structural steel tables (IS:808), we obtain

$D = 600.0$ mm, $B = 210.0$ mm, $t_w = 10.5$ mm and $t_f = 15.5$ mm

$A = 12\ 669.0\ mm^2$, $I_x = 728.7 \times 10^6\ mm^4$, $I_y = 18.2 \times 10^6\ mm^4$

where t_w = web thickness and t_f = flange thickness.

The M.I. of the section can be computed by dividing the section into three rectangular segments (top flange, web and bottom flange) as shown in Fig.6.9. The centroidal axes lie at mid-height ($\bar{y} = 300.0$ mm) and along the web central line because of the symmetry of the section. The computations for I_x and I_y are shown in Tables 6.3 and 6.4, respectively.

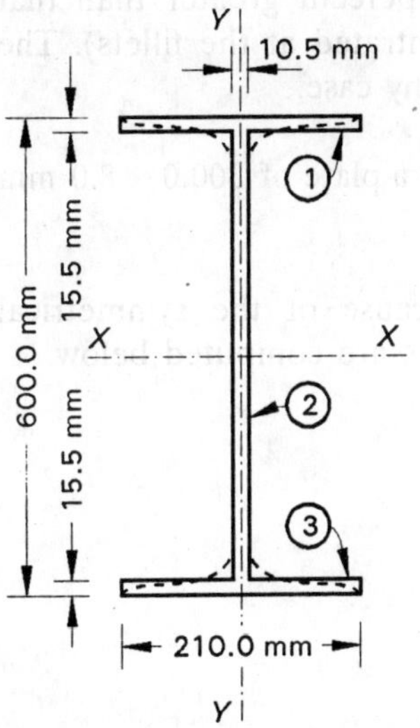

Fig.6.9 Parameters of ISLB 600 (Problem VI)

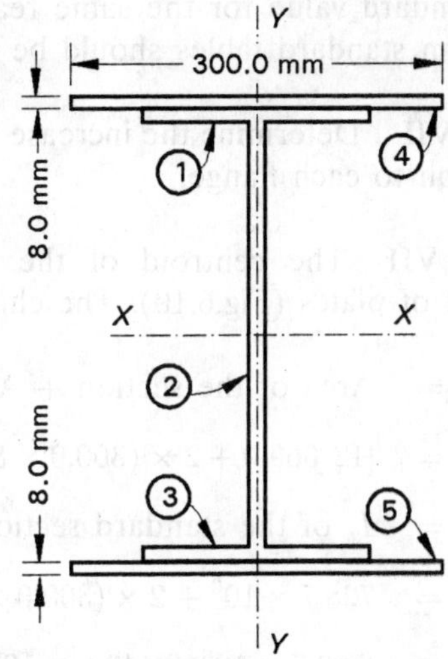

Fig.6.10 Parameters of ISLB 600 with flange plates (Problem VII)

Table 6.3 *Computations for I_x of ISLB 600 (Fig.6.9)*

Segment	a (mm^2)	y (mm)	$a(y-\bar{y})^2$ (mm^4)	I' (mm^4)
1 and 3	$2 \times (210.0 \times 15.5) = 6\,510.0$	(15.5/2)	556.02×10^6	0.13×10^6
2	$(600.0 - 2 \times 15.5) \times 10.5 = 5\,974.5$	300.0	0.00	161.19×10^6
Total	12 484.5		556.02×10^6	161.32×10^6

$$I_x = (556.02 + 161.32) \times 10^6 = 717.34 \times 10^6 \text{ mm}^4$$

Table 6.4 *Computations for I_y of ISLB 600 (Fig.6.9)*

Segment	a (mm^2)	x (mm)	$a(x-\bar{x})^2$ (mm^4)	I' (mm^4)
1 and 3	$2 \times (210.0 \times 15.5) = 6\,510.0$	(210.0/2)	0.00	23.92×10^6
2	$(600.0 - 2 \times 15.5) \times 10.5 = 5\,974.5$	(210.0/2)	0.00	0.05×10^6
Total	12 484.5		0.00	23.97×10^6

$$I_y = 23.97 \times 10^6 \text{ mm}^4$$

The values of sectional area and I_x are smaller than the values obtained from structural steel tables; the difference being 1.5 percentage for area and 1.6 percentage for I_x. The difference in the values is because of the profile of the standard sections. The actual profile comprises fillets at the web-flange junction and rounded edges as shown in Fig.6.9 in discontinuous lines. The flanges and web were assumed to be rectangular in the above computations leading to the small error. However, the computed value of I_y is about 32 percent greater than that of the standard value for the same reason (that the area is concentrated at the fillets). The values from standard tables should be adopted in the designs, in any case.

Problem VII Determine the increase in the M.I. of ISLB 600 when a plate of 300.0 × 8.0 mm is added on to each flange.

Solution VII The centroid of the section is not changed because of the symmetrical placement of plates (Fig.6.10). The changed geometrical parameters are computed below.

$$A = \text{Area of the section} + \text{Area of the plates}$$

$$= [12\ 669.0 + 2 \times (300.0 \times 8.0)]\ \text{mm}^2 = 17\ 469.0\ \text{mm}^2$$

$$I_x = I_x \text{ of the standard section} + I_x \text{ of the plates}$$

$$= 728.7 \times 10^6 + 2 \times (300.0 \times 8.0) \times (300.0 + 8.0/2)^2$$

$$= [728.7 \times 10^6 + 443.6 \times 10^6]\ \text{mm}^4 = 1\ 172.3 \times 10^6\ \text{mm}^4$$

$$I_y = I_y \text{ of the standard section} + I_y \text{ of the plates}$$

$$= 18.2 \times 10^6 + 2 \times (8.0 \times 300.0^3/12)$$

$$= [18.2 \times 10^6 + 36.0 \times 10^6]\ \text{mm}^4 = 54.2 \times 10^6\ \text{mm}^4$$

It can be noted that the area of the section increases by 37.9 percent, while I_x increases by 60.9 percent and I_y increases by 197.8 percent by the addition of flange plates. Students are advised to note the contribution of various segments of the section to the second moments of area. Such sections, comprising standard steel sections and flange plates (or other standard sections such as angles and channels), are known as *built-up sections*; these are widely used in structures.

6.5 Section modulus

It should be noted that the sectional profile, rather than the cross-sectional area, influences the moment carrying capacity of a beam significantly. The cross-sectional area alone does not reflect the beam capacity; the M.I. reflects only to a limited extent, since the maximum stress depends not only upon the M.I. but also upon the distance of the farthest fibre from the neutral axis. The maximum stress in a section can be obtained from Eqn.6.5; we have

$$\sigma_{\max} = M\, y_{\max}/I \qquad [6.11]$$

The above equation can also be written as

$$M = \sigma_{\max} [I/y_{\max}] = \sigma_{\max} Z \quad [6.12]$$

The parameter Z depends upon the sectional properties, and is known as the *modulus of the section* (or *section modulus*). The above equation indicates that larger the sectional modulus the larger will be the moment carrying capacity of the section for a given material. It is also known as the *elastic modulus* of a section, since it pertains to the stresses within the elastic limit of the material. The reference axis is also mentioned in order to avoid ambiguity; Z_{xx} (or Z_x) indicates the section modulus about the X-axis, and Z_{yy} (or Z_y) indicates the section modulus about the Y-axis.

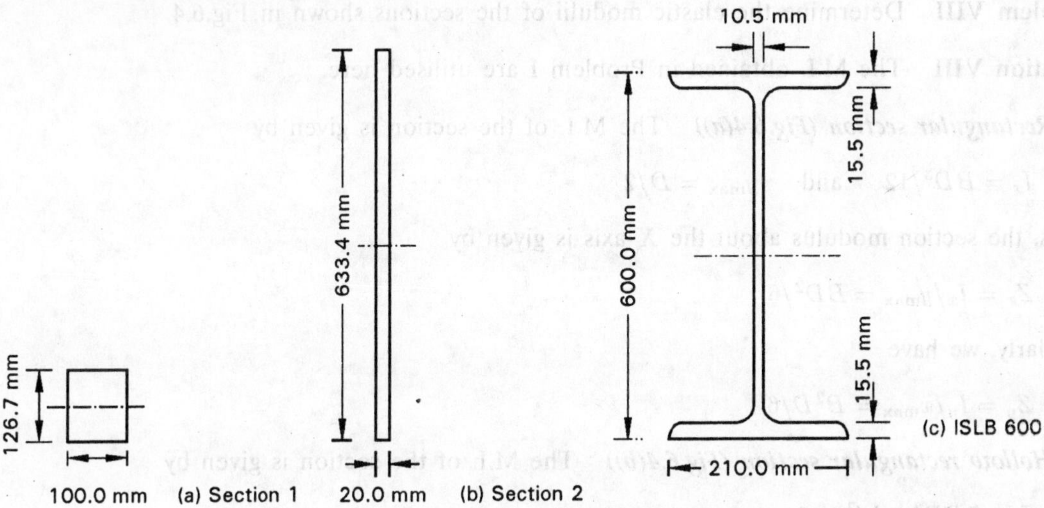

Fig.6.11 Comparison of sectional properties

The M.I. and section modulii of two rectangular sections and an I-section of the same cross-sectional area are compared for illustration in Fig.6.11. The area of ISLB 600 is 12 669.0 mm^2 (Problem VI). Consider two rectangular sections of size 100 × 126.69 mm and 20.0 × 633.45 mm; these have the same area as ISLB 600 (namely, 12 669.0 mm^2). However, the M.I. are in the ratios

Section 1 (100.0 × 126.69) : Section 2 (20.0 × 633.45) : ISLB 600

$$= 16.9 \times 10^6 \;:\; 423.6 \times 10^6 \;:\; 728.7 \times 10^6 = 1 \;:\; 25.0 \;:\; 43.0$$

The modulii of the sections are in the ratio

Section 1 : Section 2 : ISLB 600

$$= 0.27 \times 10^6 \;:\; 1.34 \times 10^6 \;:\; 2.43 \times 10^6 = 1 \;:\; 5.0 \;:\; 9.0$$

The moment carrying capacities are in the same ratio as that of the section modulii for the given material (Eqn.6.12). Though of the same cross-sectional areas, the moment carrying capacity of Section 2 (20.0 × 633.45 mm) is five times that of Section 1 (100.0 × 126.69 mm), while the capacity of ISLB 600 is nine times that of Section 1 and nearly twice that of Section 2. In other words, ISLB is the most efficient of the three sections in resisting the applied moments.

It is advantageous to concentrate the area of section at the extreme fibres in order to obtain a large capacity in bending. However, the web should be of adequate strength (thickness) to sustain the S.F. that usually accompanies bending in beams. Further, thin rectangular sections are not practical; Section 2, for instance, fails by buckling (bending in the transverse direction), and cannot be used as a beam section.

Standard steel sections (IS:808) are developed keeping in view the practical considerations discussed above so that the material is utilised in the best possible manner.

Problem VIII Determine the elastic modulii of the sections shown in Fig.6.4.

Solution VIII The M.I. obtained in Problem I are utilised here.

(a) Rectangular section (Fig.6.4(a)) The M.I. of the section is given by

$$I_x = BD^3/12 \quad \text{and} \quad y_{\max} = D/2$$

Thus, the section modulus about the X-axis is given by

$$Z_x = I_x/y_{\max} = BD^2/6$$

Similarly, we have

$$Z_y = I_y/x_{\max} = B^2D/6$$

(b) Hollow rectangular section (Fig.6.4(b)) The M.I. of the section is given by

$$I_x = (BD^3 - bd^3)/12$$

and $y_{\max} = D/2$ (the distance of the farthest fibre from the neutral axis). The modulus of the section is, thus

$$Z_x = (BD^3 - bd^3)/(6D)$$

Similarly, we have

$$Z_y = (B^3D - b^3d)/(6B)$$

(c) Triangular section (Fig.6.4(c)) The M.I. about the neutral axis is given by

$$I_x = BD^3/36 \quad \text{and} \quad y_{\max} = 2D/3$$

$$Z_x = BD^2/24$$

(d) Circular section (Fig.6.4(d)) The M.I. of the section is given by

$$I_x = \pi D^4/64 \quad \text{and} \quad y_{\max} = D/2.$$

Thus, we have

$$Z_x = \pi D^3/32 = \pi R^3/2$$

6.6 Beams of uniform strength

Bending moment is not always constant along the span of a beam. Since the stress at a section is a function of the section modulus, the cross section of the beam can be varied to comply with BM variation. This concept is usually adopted in leaf springs on carriages in order to minimise impact loading on axles. These springs are fabricated of thin laminae of equal width and thickness but of varying lengths as indicated in Fig.6.12.

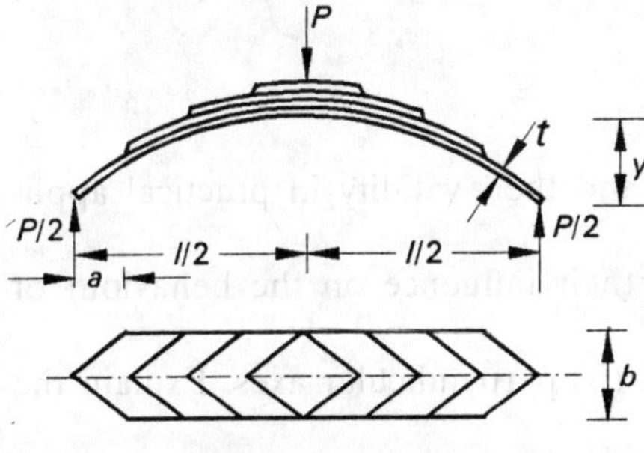

Fig.6.12 Leaf springs

Leaf springs behave like simply supported beams. The lengths of the leaves are decreased from the centre to the ends progressively to comply with the B.M. diagram. Each leaf is assumed to taper uniformly over the length by which it extends beyond the next smaller leaf. This taper is indicated in Fig.6.12; the shortest leaf is of diamond shape. The frictional force between the leaves is neglected in the analysis, and the leaves are assumed to slide over each other but maintain contact with each other. It can thus be visualised that at any section, the M.I. is the product of the M.I. of one leaf and the number of leaves at the section. The spring of n leaves is thus equivalent to a simply supported plate with a width of (nb) at the centre reducing linearly to zero at the supported ends. The M.I. (I) of the spring varies linearly along its length. The B.M. (M) also varies linearly from the maximum value at the centre, where the spring thickness is the maximum, to zero at the ends. The spring thus can be said to have constant (M/I) over its length; in other words, the stress is constant over the spring length.

6.7 Summary

Transverse loads on beams induce B.M. and S.F.; these effects were discussed in Chapter 5. The B.M. and S.F. cause stresses in the cross section of a beam. The development of simple bending theory and its applications are discussed in this chapter. The assumptions of the simple bending theory, known as Euler–Bernoulli's bending theory, are explained. The concepts of the M.I. of a plane figure (cross section of a beam), and the related theorems are presented. Beams of typical cross sections, including structural steel sections, are analysed.

The concept of section modulus is explained. The formulations of the theory are summarised below:

Euler–Bernoulli's bending equation : $[M/I] = [\sigma/y] = [E/R]$

Moment of inertia of a section about an axis : $I = \int_A y^2\, dA$

Theorem of parallel axes : $I_a = I_b + A\,a^2$

Theorem of perpendicular axes : $I_z = I_x + I_y$

Section modulus : $Z = I/y_{\max}$

6.8 Exercises

6.1 Discuss the assumptions of simple bending theory, and their validity in practical applications.

6.2 Define M.I. and modulus of a section, and discuss their influence on the behaviour of beams.

6.3 State and prove the theorem of (i) parallel axes and (ii) perpendicular axes. Explain the significance of these theorems.

6.4 Show that a square section is more efficient for a beam than a circular section of the same cross-sectional area.

6.5 Determine the moment of inertia and section modulus of a square section of side a about its diagonal.

6.6 A beam section comprises ISMC 250 placed over ISWB 450 such that the flange of the latter is inside the channel section. Determine the increase in the resistance of the section compared to ISWB 450. [Dimensions of ISWB 450: b_f=200.0 mm, t_f=15.4 mm and t_w= 9.2 mm. Dimensions of ISMC 250: b_f= 80.0 mm, t_f= 14.1 mm and t_w = 7.1 mm.]

6.7 Determine the change in the moment capacity of ISWB 450, when a 300.0 × 12.0 mm plate is welded to its (i) top flange and (ii) top and bottom flanges.

6.8 Determine whether it is economical to use two ISWB 450 sections or one ISWB 450 with 300.0 mm wide flange plates.

6.9 Compute the stresses in a simple beam of ISWB 450 section when supporting a u.d.l. of 27.0 kN/m over 6.3 m span.

6.10 A T-beam of 1.0 m overall depth, 0.25 m thick web, and a flange of 1.25 m width and 0.12 m thickness supports a load of 60.0 kN/m as a contilever of 2.7 m span. Determine the stress distribution, and estimate the maximum load it can carry, if the allowable stress is 15.0 MPa in compression and 1.5 MPa in tension.

6.11 Determine the moment carrying capacity of a simple beam of I-section of 1.25 m overall depth, 0.30 m thick web, top flange of 1.50 m width and 0.10 m thickness, and bottom flange of 0.60 m width and 0.50 m depth, if the allowable stresses are 10.0 MPa in compression and 1.0 MPa in tension.

6.12 Compute the maximum span over which an ISWB 450 section can support a udl of 50.0 kN/m for an allowable stress of 250.0 MPa as a (i) cantilever, and (ii) simple beam.

7 Combined stresses

7.0 Introduction

Structural members analysed so far were subjected to either direct forces or transverse loads causing bending. In many a case, structural elements are subjected to the combined influence of A.F. and bending. The combination of A.F. and bending may occur due to eccentricity of loading, variation of cross section along length or axial and transverse forces applied together. It should be noted that, in general, concentric loading seldom occurs in practice, and some eccentricity of axial loads is inevitable. Some examples of combined axial and transverse loadings are tall structures (transmission poles, television towers, bridge piers, high rise buildings and chimneys) subjected to wind and seismic loading. Retaining walls and hydraulic structures (dams) have to sustain gravity loads as well as transverse loads. Columns and truss members are usually subjected to eccentric axial loads as well as B.M.

The analysis of such structural elements is discussed in this chapter. It was assumed in the bending theory (Chapter 6) that the cross section was free from forces other than B.M. The influence of direct forces and flexure can be computed separately and added together; the material is assumed to be of linear elasticity, so that the principle of superposition can be applied. Several cases of members with combined loading are presented and the application of the principle of superposition is explained.

7.1 Analysis

Consider a column of arbitrary cross section subjected to a compressive force at an eccentricity e_x (the subscript indicates the co-ordinate axis along which the load is eccentric). The cross section of the member is also indicated in Fig.7.1. The eccentric load is equivalent to an axial (concentric) force P and a moment $P e_x$ about the Y-axis. The influence of these two forces can be considered separately; the A.F. causes a uniform stress, and the moment induces flexural stresses as indicated in Fig.7.1.

7.1.1 Eccentricity in one direction

Since the material is assumed to be linearly elastic, applying the principle of superposition, we obtain

$$\sigma = -(P/A) - (M_y\, x/I_y) \qquad [7.1]$$

where $M_y = P e_x$ and e_x = eccentricty along the X-axis. The subscripts in the above equation indicate the corresponding axes. The above equation is of general validity and does not place any restrictions on the shape of cross section or eccentricty of loads. If the load is eccentric about the Y-axis, Eqn.7.1 can be modified suitably; we have

$$\sigma = -(P/A) - (M_x\, y/I_x) \qquad [7.2]$$

where $M_x = P e_y$ and e_y = eccentricty along the Y-axis.

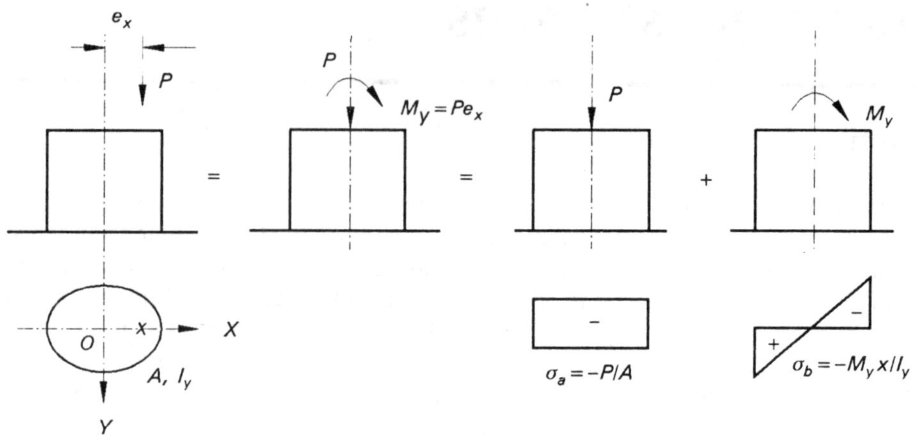

Fig.7.1 Eccentric loading on an arbitrary section

7.1.2 Eccentricity in two directions

The above equations can be combined to include eccentricties of the load about both the axes; we have

$$\sigma = -(P/A) - (M_y x/I_y) - (M_x y/I_x) \quad [7.3]$$

The application of the above equations is illustrated with the help of a few examples.

Problem I Determine the stress distribution on a rectangular section subjected to an eccentric loading.

Solution I Consider a rectangular section of width B and depth D as indicated in Fig.7.2(a); the load is eccentric along the Y-axis. We have for a rectangular section

$$A = BD, \quad y_{\max} = (D/2) \quad \text{and} \quad I_x = BD^3/12$$

Applying Eqn.7.2, and noting the negative eccentricity of the applied load, we obtain

$$\begin{aligned} \sigma &= -(P/A) - [P(-e_y)y/I_x) \\ &= -[P/(BD)] + [12Pe_y y/(BD^3)] \end{aligned} \quad [I.1]$$

The first component of the above stress condition causes uniformly distributed stress, while the second part causes linearly varying stress as indicated in Fig.7.2(b). The extreme fibre stresses can be computed from the above equation by substituting the corresponding values of y; the maximum stress occurs at $y = +D/2$, and the minimum stress occurs at $y = -D/2$.

Thus, we have

$$\sigma_{max} = -[P/(BD)] + [P\,e_y/(BD^2/6)]$$

$$\sigma_{min} = -[P/(BD)] - [P\,e_y/(BD^2/6)] \qquad [I.2]$$

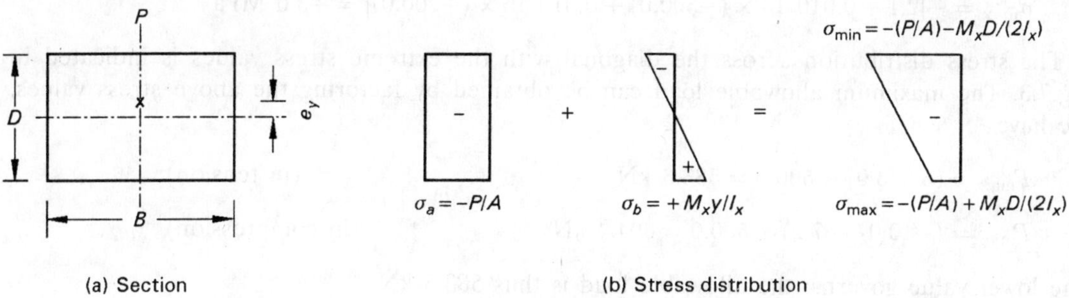

Fig.7.2 Eccentric loading on a rectangular section (Problem I)

Problem II Determine the extreme stresses on a rectangular section of size 600.0 × 400.0 mm, when a load of 500.0 kN is applied at 154.0 mm from a short edge and 135.0 mm from a long edge. Determine the maximum allowable load on the section, if the allowable stresses for the material are + 3.5 MPa and − 10.0 MPa.

Solution II The load position from the centroid of the section is noted in Fig.7.3. For the given data, we have

$$e_x = (600.0/2) - 154.0 = 146.0 \text{ mm}; \qquad e_y = (400.0/2) - 135.0 = 65.0 \text{ mm}$$

$$M_x = P\,e_y = (500.0 \times 10^3) \times 65.0 = 32.5 \times 10^6 \text{ Nmm}$$

$$M_y = P\,e_x = (500.0 \times 10^3) \times 146.0 = 73.0 \times 10^6 \text{ Nmm}$$

$$A = 600.0 \times 400.0 = 0.24 \times 10^6 \text{ mm}^2$$

$$I_x = 600.0 \times 400^3/12 = 3.2 \times 10^9 \text{ mm}^4; \quad I_y = 400.0 \times 600^3/12 = 7.2 \times 10^9 \text{ mm}^4$$

Substituting the above values in Eqn.7.3, we obtain

$$\sigma = -(P/A) - (M_y\,x/I_y) - (M_x\,y/I_x)$$

$$= -[500.0 \times 10^3/(0.24 \times 10^6)] - [73.0 \times 10^6\,x/(7.2 \times 10^9)]$$

$$-[32.5 \times 10^6\,y/(3.2 \times 10^9)] \qquad [x \text{ and } y \text{ in mm}]$$

$$= -[2.1 + 0.010\,14\,x + 0.010\,16\,y] \text{ MPa} \qquad [II.1]$$

It is easy to note that minimum stress occurs at the corner closest to the applied load (+ 300.0, + 200.0), while the maximum stress occurs at the farthest corner (– 300.0, – 200.0). Accordingly, we have

$$\sigma_{min} = -[2.1 + 0.010\ 14 \times 300.0 + 0.010\ 16 \times 200.0] = -7.2 \text{ MPa}$$

$$\sigma_{max} = -[2.1 + 0.010\ 14 \times (-300.0) + 0.010\ 16 \times (-200.0)] = +3.0 \text{ MPa}$$

The stress distribution across the diagonal with the extreme stress values is indicated in Fig.7.3. The maximum allowable load can be obtained by factoring the above stress values. We have

$$P_{max} = (3.5/3.0) \times 500.0 = 583.3 \text{ kN} \qquad \text{(in tension)}$$

$$P_{max} = (-10.0/-7.2) \times 500.0 = 694.4 \text{ kN} \qquad \text{(in compression)}$$

The lower value governs; the allowable load is thus 583.3 kN.

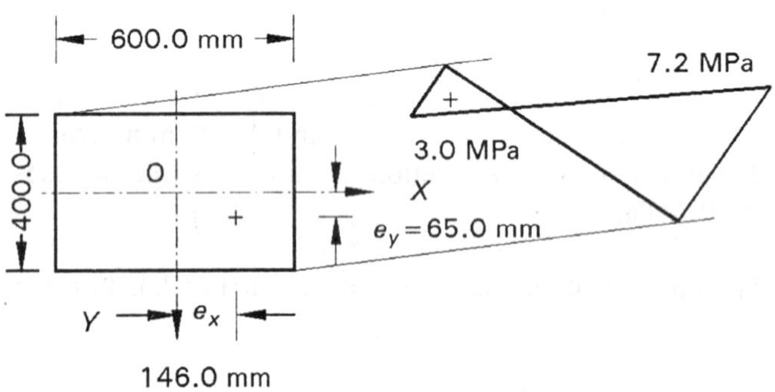

Fig.7.3 Stress in a rectangular section (Problem II)

7.2 Kern of a section

The combination of axial and bending stresses induces linearly varying stresses on the cross section. For some combination of compressive A.F. and B.M., the stresses may be tensile. Massive structures such as dams, abutments and retaining walls are often constructed in masonry. Masonry structures cannot sustain tensile stresses, and are proportioned to avoid tensile stresses in any part of the structures. Referring to Fig.7.1, if tensile stresses are not permissible, we have the condition that $\sigma_{max} \leq 0$

or $-(P/A) + (P\, e_x\, x_e/I_y) \leq 0$ [x_e = extreme fibre distance]

Simplifying the above equation, we obtain

$$e_x \leq I_y/(x_e\, A) \qquad [7.4(a)]$$

Similarly, for load eccentricity along the Y-axis, we have (y_e = extreme fibre distance)

$$e_y \leq I_x/(y_e\, A) \qquad [7.4(b)]$$

The section does not develop any tensile stresses when the eccentricities are within the limits given by the above equations. The eccentricities along the negative axes can be obtained by substituting the maximum fibre distances in these equations. The influence of the eccentricities along both the axes can be investigated by considering Eqn.7.3. From the condition of zero tensile stresses we have

$$-(P/A) + (P\,e_x\,x_e/I_y) + (P\,e_y\,y_e/I_x) \leq 0$$

Simplifying the above equation, we obtain

$$(e_x\,x_e\,A/I_y) + (e_y\,y_e\,A/I_x) \leq 1.0 \qquad [7.5]$$

Since the M.I. and extreme fibre distances are constant for a given cross section, the above equation is linear for the parameters e_x and e_y. The region defined by the equation is bounded by straight lines around the centroid of the section. The stresses on the cross section remain compressive, so long as the eccentricities lie within the region defined by Eqn.7.5. This region is known as the *kern* of the section; sometimes, it is also known as the *core* of the section. The extent of kern depends upon the section profile (sectional properties, and extreme fibre distances). The kern of a few typical geometric figures is determined here.

7.2.1 Rectangular section (middle third rule)

The geometrical properties of a rectangular section (Fig.7.2) are substituted in Eqn.7.5 in order to determine the kern of the section. For the section shown in Fig.7.4, we have

$$A = BD, \quad I_x = BD^3/12, \quad \text{and} \quad I_y = B^3D/12$$

$$x_e = \pm B/2 \quad \text{and} \quad y_e = \pm D/2$$

Substituting in Eqn.7.5 and simplifying, we obtain

$$\pm(6\,e_x/B) \pm (6\,e_y/D) = 1.0 \qquad [7.6]$$

The extreme values of e_x (eccentricity along the X-axis) are according to the above equation $\pm(B/6)$ along the X-axis, and those of e_y are $\pm(D/6)$. Between these extreme values, the eccentricities have linear variation as indicated in Fig.7.4. The parallelogram bound by the middle third points of a rectangular section is its kern; this observation is known as the *middle third rule.*

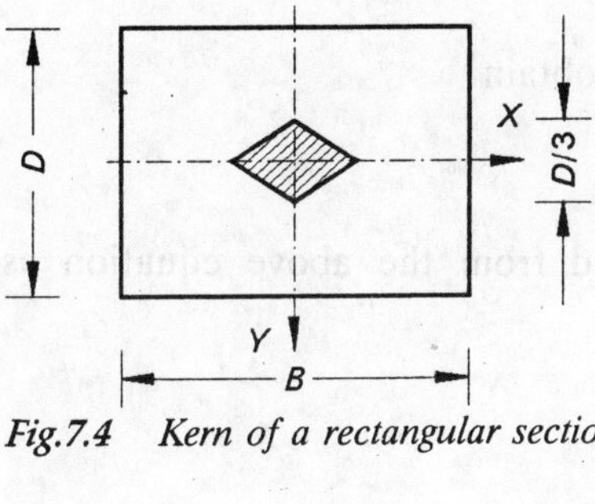

Fig.7.4 Kern of a rectangular section

7.2.2 Circular section (middle quarter rule)

For a circular section of diameter D, we have

$$A = \pi D^2/4, \quad I_y = \pi D^4/64 \quad \text{and} \quad x_e = D/2$$

Because of the symmetry of the section along any radial line, we obtain from Eqn.7.4(a)

$$e_x \leq \frac{I_y}{(x_e A)} \leq \frac{(\pi D^4/64)}{[(D/2) \times D^2/4]} \leq \frac{D}{8} \qquad [7.7]$$

The extreme values of the eccentricity along any radial line are, according to the above equation, $\pm(D/8)$. The kern of a circular section is thus a circle of diameter equal to a quarter of the diameter of the section as shown in Fig.7.5; this is known as the *middle quarter rule*.

Equation 7.5 is of general validity, and can be applied to any cross section subjected to eccentric loading. The application of the equation is illustrated through a typical example.

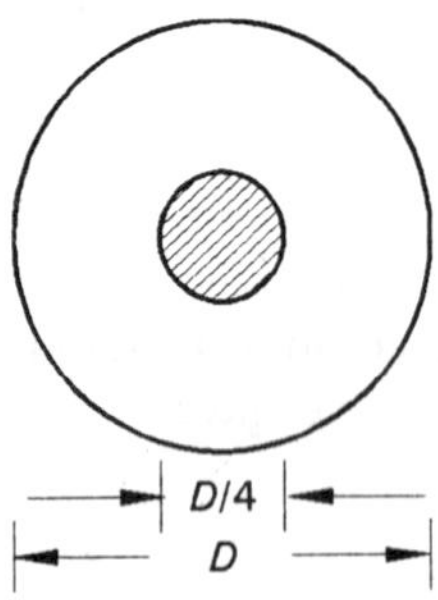

Fig.7.5 Kern of a circular section

Problem III Determine the kern of ISLB 600.

Solution III The properties of the section are adopted from Problem VI, Chapter 6; we have

$$A = 12\ 484.5 \text{ mm}^2, \quad I_x = 717.34 \times 10^9 \text{ mm}^4 \cdot$$

$$I_y = 23.97 \times 10^6 \text{ mm}^4, \quad y_e = 300.0 \text{ mm}$$

and $x_e = 105.0$ mm

Substituting the above values in Eqn.7.5 and simplifying, we obtain

$$0.054\ 7 e_x + 0.005\ 2 e_y = 1.0$$

The extreme values of the eccentricities can be determined from the above equation as (because of the symmetry of the section)

$$e_x = \pm 18.28 \text{ mm} \quad \text{and} \quad e_y = +192.31 \text{ mm}$$

The kern of the section is a parallelogram with the extreme values of ±18.28 mm on the X-axis and ±192.31 on the Y-axis. The kern is shown in Fig.7.6.

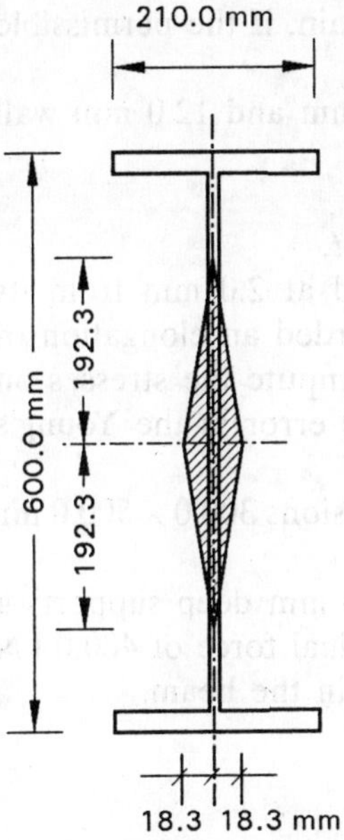

Fig.7.6 Kern of ISLB 600 (Problem III)

7.3 Summary

Analysis of structural elements subjected to the combined effects of direct loading and B.M. is presented. The concepts of eccentric loading and kern of a section are explained. Various types of cross sections are analysed.

The stress in a section subjected to eccentric loading is given by

$$\sigma = -(P/A) - (M_y\, x/I_y) - (M_x\, y/I_x)$$

The condition for zero tensile stresses in a cross section is given by

$$(e_x\, x\, A/I_y) + (e_y\, y\, A/I_x) \leq 1.0$$

7.4 Exercises

7.1 Develop the stress conditions for a member subjected to A.F. when (i) the load is eccentric along the X-axis and (ii) the load is eccentric along both the axes.

7.2 Define the kern of a section, and discuss its significance in structural analysis.

7.3 What is the middle third rule? Explain its significance.

7.4 What is the middle quarter rule? Explain its significance.

7.5 A rectangular column of size 250.0 × 350.0 mm is subjected to a load of 500.0 kN at 120.0 mm from a shorter edge and 45.0 mm from the adjacent longer edge. Determine the stress distribution. What is the maximum load the column can sustain, if the permissible stresses are + 3.5 MPa and − 35.0 MPa?

7.6 Determine the kern of a circular section of external radius 250.0 mm and 12.0 mm wall thickness.

7.7 Determine the kern of a thin cylindrical section.

7.8 Determine the kern of a thin square tube of side a and thickness t.

7.9 A 20.0 mm circular bar is tested in tension with the load applied at 2.0 mm from its centre. An extensometer fixed on the line of the applied load recorded an elongation of 0.18 mm over a gage length of 120.0 mm at a load of 40.0 kN. Compute the stresses on the section and the Young's modulus of the material. Estimate the error in the Young's modulus, if the eccentricity is ignored.

7.10 Determine the kern of a hollow rectangular section of external dimensions 300.0 × 500.0 mm and 75.0 mm wall thickness.

7.11 A cantilever beam of rectangular section 250.0 mm wide and 400.0 mm deep supports a vertical load of 50.0 kN at 2.5 m from the fixed end and a longitudinal force of 400.0 kN at 50.0 mm below the top fibre. Determine the stress distribution in the beam.

8 Torsional stresses

8.0 Introduction

Basic concepts of torsion were explained in Chapter I, and illustrated in Figs.I.4 and I.5. Torsion occurs in several systems, such as three-dimensional (space) structures, machine elements, aircraft structures, ships, shafts transmitting power, springs, beams, box girders, slabs, shear walls and frames. Often, torsional moments occur in combination with B.M. and A.F. Torsional effects on circular members are different from those on non-circular members. Torsion causes a member to twist, and by implication induces shearing stresses (Chapter 3). A non-circular member subjected to torsion develops cross-sectional distortion besides shear stresses and deformations, while circular members are free from torsional distortion.

The analysis of circular shafts subjected to torsional moments is discussed in this chapter. The concepts of torsional stresses, deformations and strain energy are explained; members with varying torque and compound shafts are also discussed.

8.1 Torsion analysis

Torsional behaviour of circular members is much different from that of non-circular members. The analysis of circular members is discussed here; the analysis can be extended to other cross sections. The assumptions of basic torsional analysis are explained, and the torsional equation, similar to the bending equation is developed.

8.1.1 Assumptions of torsion analysis

A circular shaft subjected to pure torsion develops shear stresses in planes normal to the axis of torsion (polar axis) as indicated in Fig.8.1. The shear stresses occur in the direction perpendicular to the radial lines of shaft cross section; complementary shear stresses are

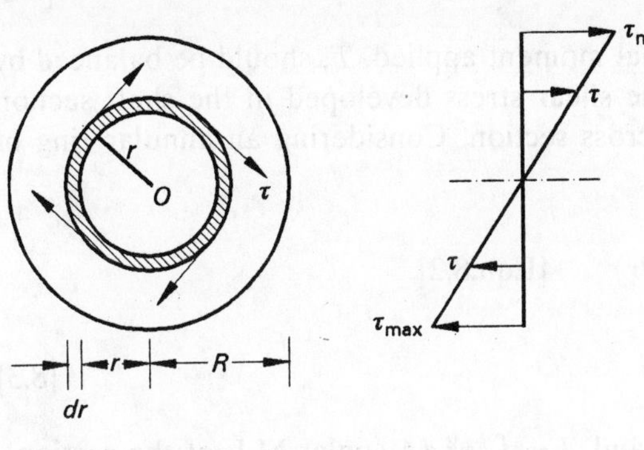

Fig.8.1 Torsional stresses

developed in the longitudinal direction of the shaft. The stresses in an elemental ring of the shaft cross section are indicated in the figure.

The analysis of members subjected to pure torsion is based on the simplifying assumptions listed below:

1. The material is an isotropic homogeneous elastic continuum.
2. Sections plane before twisting remain plane after twisting.
3. The shaft is straight and of uniform section.
4. Strains and deformations are small.

It may be noted that these assumptions are similar to those of bending theory. Assumption 2 implies that strains do not occur in the direction of the polar axis of the member; this is valid only for circular shafts. In non-circular shafts, plane sections do not remain plane after twisting; this effect is known as *warping* of cross section. Warping stresses are significant in box girders, standard steel sections, and aircraft structures. These stresses are not considered in the analyses presented here, however.

8.1.2 Torsion equation

Consider a circular shaft of length l and radius R indicated in Fig.8.2 subjected to pure torsion. The shaft is assumed fixed at one of the ends and free to rotate at the other end when subjected to a twisting moment. The twisting of the shaft displaces radial fibre OA to OB through a small angle θ (A and B are at a radial distance r from polar axis O). From the condition of compatibility, the displacement of A in the cross section and longitudinal section should be the same. If the longitudinal line CA is rotated through ϕ to CB due to torsion, we have

$$r\,\theta = l\,\phi \qquad [8.1(a)]$$

However, the angular change in the longitudinal direction is shear strain by definition. If the shear stress in the shaft is τ, and the material has a rigidity modulus G, we have

$$\phi = \tau/G \qquad [8.1(b)]$$

Combining the above two equations, we obtain

$$[\tau/r] = [G\,\theta/l] \qquad [8.2]$$

Applying the equation of equilibrium, the torsional moment applied, T, should be balanced by the torsional resistance of the shaft. Let τ be the shear stress developed in the shaft section at a radial distance r from the centroid of the cross section. Considering an annular ring of thickness dr (Fig.8.1), we have

$$T = \int_A (2\pi\,r\,dr)\tau\,r = [G\theta/l]\int_A r^2(2\pi\,r\,dr) \qquad \text{[Eqn.8.2]}$$

$$= [G\theta/l]\int_A r^2 dA = [G\,\theta/l]\,J \qquad [8.3]$$

where $dA = 2\pi\,r\,dr$ (area of the elemental ring) and $J = \int_A r^2\,dA$ (polar M.I. of the section).

Combining Eqns.8.2 and 8.3, we obtain

$$\frac{T}{J} = \frac{G\theta}{l} = \frac{\tau}{r} \qquad [8.4]$$

The above equation is analogous to the bending equation (Eqn.6.5). It indicates that shear stress at any point in a shaft is proportional to its distance from the polar axis and to the applied torque. That is, shear stress varies linearly from zero at the shaft centre to the maximum value at the periphery of the section. The parameter (θ/l) indicates the rotation of the shaft per unit length, and is analogous to the curvature of a beam. The parameter GJ is known as the *torsional rigidity* of the shaft (similar to EI, the flexural rigidity). *Torsional stiffness* of a shaft is defined as the torsional moment per unit angle of twist (in radians); mathematically

$$\text{Torsional stiffness} \quad k = [T/\theta] = [GJ/l] \qquad [8.5(a)]$$

The maximum shear stress in the section occurs at the periphery $(r = R)$; we have

$$\tau_{max} = T\,R/J = T/C \qquad [8.5(b)]$$

where $C = [J/R]$. The parameter C is known as *torsional modulus* of the section similar to Z, the section modulus (Eqn.6.12).

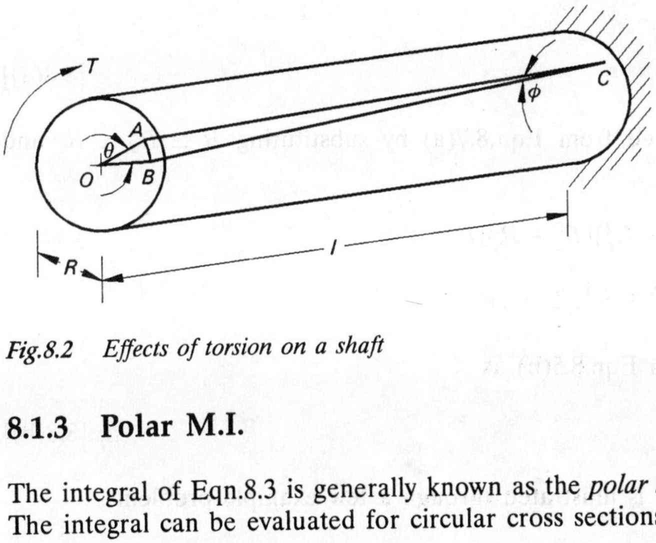

Fig.8.2 Effects of torsion on a shaft

8.1.3 Polar M.I.

The integral of Eqn.8.3 is generally known as the *polar moment of inertia* of the shaft section. The integral can be evaluated for circular cross sections as indicated below.

8.1.3.1 Solid circular section

Referring to Fig.8.3(a), for a section of radius R, we have

$$J = (2\pi)\int_0^R r^3\,dr \quad [dA = 2\pi\,r\,dr]$$

$$= \pi\,R^4/2 \qquad [8.6(a)]$$

It may be recalled that the above relation was also developed in Chapter 6 (Problem I). The maximum shear stress occurs at the periphery of the section (at $r = R$); we have

$$\tau_{max} = T\,R/J = 2T/(\pi\,R^3) \qquad [8.6(b)]$$

8.1.3.2 *Hollow circular section*

The integral of Eqn.8.3 is to be evaluated for the limits of inner radius R_i and outer radius R_o of a hollow circular cylindrical section; accordingly, we have (Fig.8.3(b))

$$J = (2\pi) \int_{R_i}^{R_o} r^3\,dr = \pi\,(R_o^4 - R_i^4)/2 \qquad [8.7(a)]$$

The maximum shear stress occurs at the periphery of the section (at $r = R_o$); we have

$$\tau_{max} = 2T\,R_o/[\pi\,(R_o^4 - R_i^4)] \qquad [8.7(b)]$$

8.1.3.3 *Thin circular tube*

For a circular tube of mean radius R and wall thickness t, the polar M.I. can be expressed as (Fig.8.3(c))

$$J = (2\pi\,R\,t)\,R^2 = 2\,\pi\,R^3\,t \qquad [8.8(a)]$$

The above value can also be obtained from Eqn.8.7(a) by substituting $R \simeq R_o \simeq R_i$ and $t = (R_o - R_i)$; we have

$$\begin{aligned} J &= \pi\,(R_o^4 - R_i^4)/2 = \pi\,(R_o^2 - R_i^2)(R_o^2 + R_i^2)/2 \\ &= \pi\,(R_o - R_i)(R_o + R_i)(R_o^2 + R_i^2)/2 = 2\,\pi\,R^3\,t \end{aligned}$$

The maximum stress is obtained from Eqn.8.5(b) as

$$\tau_{max} = T/(2\pi\,R^2\,t) \qquad [8.8(b)]$$

The application of the above analysis is illustrated through a few example problems.

Problem I Determine the maximum stress and deformation of a shaft of 100.0 mm diameter and 2.7 m length subjected to a torque of 30.0 kNm. Assume $G = 75.0$ GPa for the material.

Solution I The polar M.I. of the shaft is given by

$$\begin{aligned} J &= \pi\,R^4/2 \qquad \text{[Eqn.8.6(a)]} \\ &= \pi\,(50.0^4)/2 = 9.817 \times 10^6 \text{ mm}^4 \end{aligned}$$

Applying Eqn.8.6(b), we have

$$\tau_{max} = 2T/(\pi R^3) = 2 \times (30.0 \times 10^6)/(\pi \times 50.0^3) = 152.8 \text{ MPa}$$

The above value can be obtained by applying Eqn.8.5 as well ($\tau_{max} = TR/J$). The deformation (angle of twist) of the section can be obtained from Eqn.8.4; we have

$$\theta = Tl/(GJ) = 30.0 \times 10^6 \times 2.7 \times 10^3/[75.0 \times 10^3 \times 9.817 \times 10^6]$$

$$= 0.110 \text{ radians} \quad [= 6.3°]$$

Problem II Determine the maximum stress and deformation of a shaft of the previous problem for a cross section of 100.0 mm external diameter and 10.0 mm wall thickness. Compare the stiffnesses of the two shafts.

Solution II From the given data, the outer and inner radii of the tube are

$R_o = 50.0$ mm and $R_i = 40.0$ mm

The polar M.I. of the shaft is given by

$$J = \pi (R_o^4 - R_i^4)/2 \quad \text{[Eqn.8.7(a)]}$$

$$= \pi (50.0^4 - 40.0^4)/2 = 5.796 \times 10^6 \text{ mm}^4$$

Applying Eqn.8.7(b), we have

$$\tau_{max} = 2T R_o/[\pi (R_o^4 - R_i^4)] = 2 \times (30.0 \times 10^6) \times 50.0/[\pi (50.0^4 - 40.0^4)]$$

$$= 258.8 \text{ MPa}$$

The deformation (angle of twist) of the section can be obtained from Eqn.8.4; we have

$$\theta = Tl/(GJ) = 30.0 \times 10^6 \times 2.7 \times 10^3/[75.0 \times 10^3 \times 5.796 \times 10^6]$$

$$= 0.186\,3 \text{ radians} \quad [= 10.7°]$$

From Eqn.8.5, we have ratio of the stiffnesses of the shafts

$$= (GJ/l)_{solid}/(GJ/l)_{tube} = (J)_{solid}/(J)_{tube}$$

$$= (9.871 \times 10^6)/(5.796 \times 10^6) = 1.703$$

The solid shaft is thus 1.7 times stiffer than the hollow tube but has about 2.8 times the area of the tube.

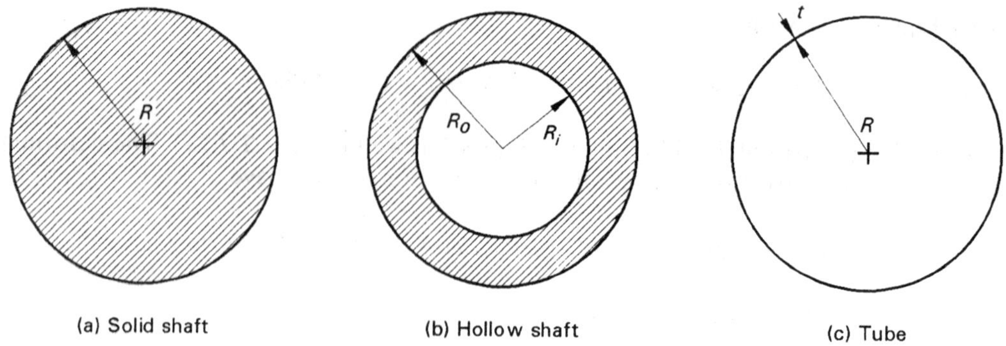

Fig.8.3 Polar M.I.

Problem III A shaft of 11.0 m length fixed at both the ends is subjected to an anti-clockwise torque of 45.0 kNm at a distance of 4.5 m from one end, and a clockwise torque of 63.0 kNm at 3.0 m from the other end. Determine the minimum size of the shaft, if the allowable shear stress is 95.0 MPa. Compute the maximum twist and the section of zero twist in the shaft assuming $G = 75.0$ GPa.

Solution III Referring to Fig.8.4, the torque at A is assumed to be T. The torques in segments AC, CD and DB can be computed from the equations of equilibrium. The torque in segment AC is thus T, that in CD will be $(T - 45.0)$. The torque in segment DB will be $(T - 45.0 + 63.0)$. The torque diagram of the shaft is indicated in Fig.8.4(a). The angles of twist at ends A and B are zero; applying this criterion after noting that the angle of twist is proportional to the segment length in a prismatic shaft, we have

$$4.5\,T + 3.5\,(T - 45.0) + 3.0\,(T + 18.0) = 0 \quad \text{or} \quad T = 9.41 \text{ kNm}$$

The torques in various segments can be computed as

$$T_{AC} = 9.41 \text{ kNm}, \quad T_{CD} = 9.41 - 45.0 = -35.59 \text{ kNm}$$

$$T_{DB} = 9.41 - 45.0 + 63.0 = +27.41 \text{ kNm}$$

The maximum value of torque (numerical) in the shaft is 35.59 kNm; the shaft should be able to sustain this torque safely. We have

$$\tau_{\max} = 2T/(\pi R^3) \qquad \text{[Eqn.8.6(b)]}$$

or $$R^3 = 2T/(\pi \tau_{\max}) = 2 \times 35.59 \times 10^6/(\pi \times 95.0)$$

or $$R = 62.01 \text{ mm}$$

Note Shafts are available in standard sizes, and are listed in design data handbooks.

The minimum shaft radius required is 62.01 mm; a shaft of 130.0 mm diameter may be adopted. For this shaft, we have

$$J = \pi D^4/32 = \pi \times 130.0^4/32 = 28.04 \times 10^6 \text{ mm}^4$$

The twist in various segments can be computed by applying Eqn.8.4; we have

$$\theta_{AC} = [T\,l]_{AC}/(GJ) = 9.41 \times 10^6 \times 4.5 \times 10^3/(75.0 \times 10^3 \times 28.04 \times 10^6)$$

$$= 0.020\ 1 \text{ radian} \qquad \text{[anti-clockwise rotation]}$$

$$\theta_{CD} = [T\,l]_{CD}/(GJ) = 35.59 \times 10^6 \times 3.5 \times 10^3/(75.0 \times 10^3 \times 28.04 \times 10^6)$$

$$= 0.059\ 2 \text{ radian} \qquad \text{[clockwise rotation]}$$

$$\theta_{DB} = [T\,l]_{CB}/(GJ) = 27.41 \times 10^6 \times 3.0 \times 10^3/(75.0 \times 10^3 \times 28.04 \times 10^6)$$

$$= 0.039\ 1 \text{ radian} \qquad \text{[anti-clockwise rotation]}$$

The maximum rotation of 0.059 2 radian (3.392 degree) occurs in segment CD. The directions of rotations are viewed from end A of the shaft. It may be noted that the above rotations satisfy the compatibility conditions (total rotation between ends A and B is zero). The point of zero twist occurs in segment CD at a distance of 1.19 m from C (Fig.8.4(b)).

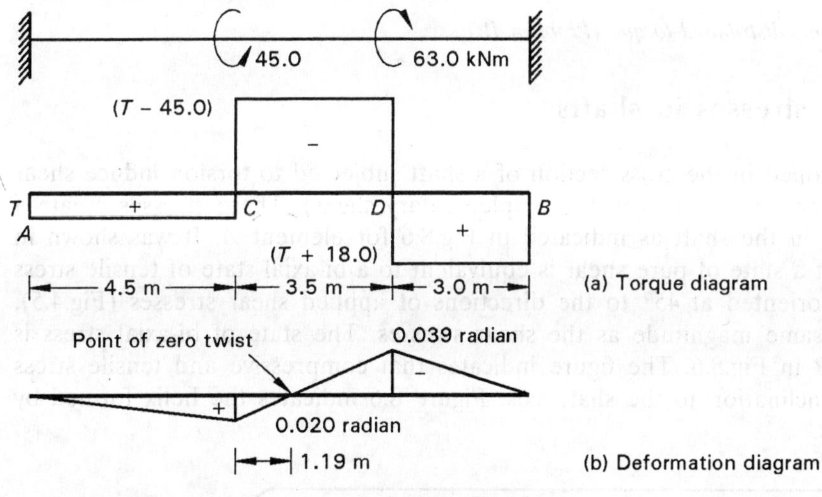

Fig.8.4 Shaft with varying torque (Problem III)

Problem IV A 3.0 m long prismatic shaft of 50.0 mm diameter, restrained at its ends, is subjected to a uniformly distributed torque of 12.6 kNm^{-1}. Determine the maximum stress and deformation of the shaft, if the rigidity modulus is 50.0 GPa.

Solution IV The torque diagram is shown in Fig.8.5. The maximum torque occurs at the supports, while the maximum deformation occurs at the mid-span section. We have

$$J = \pi D^4/32 = \pi \times 50.0^4/32 = 0.614 \times 10^6 \text{ mm}^4$$

$$\text{Maximum torque} \quad T = t\,l/2 = 12.6 \times 3.0/2 = 18.9 \text{ kNm}$$

$$\text{Maximum twist} \quad \theta = T\,(l/2)/(GJ) = 18.9 \times 10^6 \times 1.5/(50.0 \times 10^3 \times 0.614 \times 10^6)$$

$$= 0.923 \times 10^{-3} \text{ radian (0.053 degree)}$$

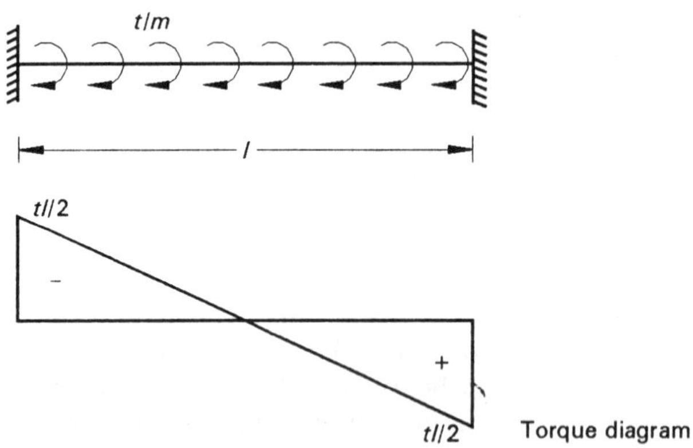

Fig.8.5 Beam with uniformly distributed torque (Problem IV)

8.1.4 Longitudinal stresses in shafts

The shear stresses developed in the cross section of a shaft subjected to torsion induce shear stresses in the longitudinal direction as well (complementary shear). These stresses create a condition of pure shear in the shaft as indicated in Fig.8.6 for element A. It was shown in Chapter 4 (Eqn.4.5) that a state of pure shear is equivalent to a bi-axial state of tensile stress and compressive stress oriented at 45° to the directions of applied shear stresses (Fig.4.5); the stresses are of the same magnitude as the shear stresses. The state of bi-axial stress is indicated for element B in Fig.8.6. The figure indicates that compressive and tensile stress trajectories run at 45° inclination to the shaft axis. Figure 8.6 indicates the helix formed by the compressive stresses.

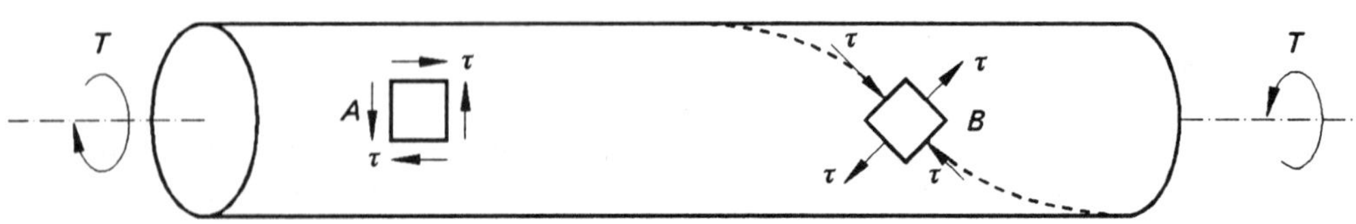

Fig.8.6 Stresses in a shaft

In a thin walled tube (Fig.8.3(c)), these compressive stresses may cause buckling (lateral deformation). This phenomenon can be demonstrated by rolling a sheet of paper into a tube and twisting it. The thickness of a tube should be at least one-sixtieth of its radius in order to avoid buckling failure.

8.2 Strain energy in elastic shafts

The concepts of work and energy were explained already in Chapter 2. The same principles can be extended to shafts subjected to torsion. The external work done by the applied torque in deforming a shaft through an angle θ is given by

$$W_e = [T\,\theta/2] \qquad [8.9(a)]$$

The shear strain energy of the shaft can be expressed as

$$U = \int dU = [1/(2G)] \int_V \tau^2\, dv \qquad [8.9(b)]$$

Considering an elemental tube of radius r and thickness dr shown in Fig.8.1, we have

$$\tau = \tau_{max}\, r/R \quad \text{and} \quad dv = (2\pi\, r\, dr)\, l = (dA)\, l$$

Substituting these parameters in Eqn.8.9(b), we have

$$U = [\tau_{max}^2\, l/(2G\,R^2)] \int_A r^2\, dA = [\tau_{max}^2\, J\,l/(2G\,R^2)] \qquad \left[J = \int_A r^2\, dA\right]$$

$$= [\tau_{max}^2/(4G)]V \qquad [\text{for solid shafts } J = \pi\,R^4/2, \text{ and } V = Al\,]$$

$$= [T^2\, l/(2G\,J)] \qquad [8.9(c)]$$

From the principle of conservation of energy, the external work done (Eqn.8.9(a)) and the strain energy (Eqn.8.9(b) or (c)) are equal ($W_e = U$). These concepts of strain energy are useful in analysing non-prismatic shafts and shafts subjected to varying torques as well as in designing shafts.

Problem V Compute the maximum torque a hollow mild steel shaft of external diameter 50.0 mm and 5.0 mm thickness can sustain with a factor of safety of 2.0, if the elastic limit in tension is 240.0 MPa. Use the criterion of (i) maximum shear stress, and (ii) maximum strain energy.

Solution V The polar M.I. of the shaft is given by

$$J = \pi\,(R_o^4 - R_i^4)/2 \qquad [\text{Eqn.8.7(a)}]$$

$$= \pi\,(25.0^4 - 20.0^4)/2 = 0.362 \times 10^6 \text{ mm}^4$$

Area of cross section $A = \pi\,(25.0^2 - 20.0^2) = 706.9 \text{ mm}^2$

Applying Eqn.8.4, the maximum shear stress in the shaft for torque T is given by

$$\tau_{max} = T\,R_o/J = T \times 25.0/(0.362 \times 10^6) \quad [T \text{ in Nmm}]$$

$$= 69.06 \times 10^{-6}\,T \qquad [\text{V.1}]$$

The limiting tensile stress in the material

$$\sigma_s = \sigma_{sy}/\text{f.o.s.} = 240.0/2.0 = 120.0 \text{ MPa}$$

Maximum allowable shear stress $\tau_{max} = \sigma_s/2 = 120.0/2 = 60.0$ MPa

Young's modulus of 200.0 GPa and Poisson's ratio of 0.30 can be assumed for mild steel. Thus, we have G = 76.92 GPa (Eqn.4.21). The maximum allowable torque can be computed for the given criteria.

(i) Maximum shear stress From Eqn.V.1, we have

$$\tau_{max} = 69.06 \times 10^{-6}\,T \quad [\tau_{max} = 60.0 \text{ MPa}] \quad \text{or} \quad T = 0.868\,8 \times 10^6 \text{ Nmm} = 868.8 \text{ Nm}$$

The maximum allowable torque based on the maximum shear stress criterion is thus 868.8 Nm.

(ii) Maximum strain energy The maximum allowable torque can be determined by equating the maximum strain energy per unit volume in torsion with that in simple tension. Applying Eqn.8.9(c), we have

$$\text{Maximum strain energy in torsion} = T^2 l/(2\,G\,J) = T^2 l/(2 \times 76.92 \times 10^3 \times 0.362 \times 10^6)$$

$$= T^2 l/(55.69 \times 10^9)$$

$$\text{Maximum strain energy in tension} = \sigma_{max}^2\,V/(2E) = 120.0^2 \times 706.9\,l/(2 \times 200.0 \times 10^3)$$

$$= 25.448\,l$$

Equating the above values of strain energies, we obtain

$$T = 1.190 \times 10^6 \text{ Nmm} = 1\,190.0 \text{ Nm}$$

The maximum allowable torque based on the maximum strain energy criterion is thus 1 190.0 Nm.

8.3 Combined torsion and bending

Torsion seldom occurs alone in practice. In many a case, structural members are subjected to torsion in combination with bending (beams and transmission shafts); axial thrust may also occur along with torsion and bending (propeller shafts).

When a section is subjected to bending, the extreme fibres are subjected to the maximum bending stresses. The maximum torsional stresses may also occur at these fibres depending upon the section configuration. A combination of the maximum bending and torsional stresses induces the maximum stress condition in the section. The principal stresses at these sections are computed using the expressions developed in Chapter 4. These aspects are discussed here briefly.

8.3.1 Solid circular shafts

If a solid circular section is subjected to a B.M. M, and a torsional moment T, the maximum stresses are given by

$$\sigma_{max} = 32\,M/(\pi\,D^3) \quad \text{and} \quad \tau_{max} = 16\,T/(\pi\,D^3)$$

The major principal stress is given by

$$\sigma_1 = (\sigma_{max}/2) + [(\sigma_{max}/2)^2 + \tau_{max}^2]^{0.5} = [16/(\pi D^3)][M + \{M^2 + T^2\}^{0.5}]$$

$$= [32/(\pi D^3)]\,M_e \qquad [8.10]$$

where $M_e = [M + \{M^2 + T^2\}^{0.5}]/2$

The equivalent B.M., M_e, induces the same maximum stresses on the section as the combination of torsional and B.M. Similarly, the maximum shear stress on the section induced by the given combination of moments is

$$\tau = [16/(\pi D^3)][M^2 + T^2]^{0.5} = [16/(\pi\,D^3)]\,T_e \qquad [8.11]$$

where $T_e = [M^2 + T^2]^{0.5}$

The parameter T_e is known as the equivalent torsion, since it causes the maximum shear stress of the same magnitude induced by the given combination of moments.

Problem VI A flywheel of 5.0 kN weight is mounted on a 70.0 mm diameter shaft mid-way between bearings 1.0 m apart. If the shaft is transmitting 50.0 kW at 450.0 rpm, determine the maximum stresses in the shaft.

Soltuion VI The B.M. in the shaft is given by

$$M = 5.0 \times 1.0/4 = 1.25 \text{ kNm}$$

The torsional moment $T = 60\,P/(2\pi\,n)$

$$= 60 \times 50.0 \times 10^3/(2 \times \pi \times 450.0) = 1\ 016.0 \text{ Nm} = 1.061 \text{ kNm}$$

Equivalent moment $M_e = [M + \{M^2 + T^2\}^{0.5}]/2$

$$= [1.25 + \{1.25^2 + 1.061^2\}^{0.5}]/2 = 1.44 \text{ kNm}$$

Equivalent torque $T_e = [M^2 + T^2]^{0.5}$

$$= [1.25^2 + 1.061^2]^{0.5} = 1.64 \text{ kNm}$$

Applying Eqn.8.10, we obtain

$$\text{The maximum bending stress} = [32/(\pi D^3)]M_e$$

$$= [32/(\pi \times 70.0^3)] \times 1.44 \times 10^6 = 42.8 \text{ MPa}$$

Applying Eqn.8.11, we obtain

$$\text{The maximum shear stress} = [16/(\pi D^3)]T_e = [16/(\pi \times 70.0^3)] \times 1.64 \times 10^6$$

$$= 24.4 \text{ MPa}$$

Problem VII A propeller shaft of 100.0 mm external diameter and 60.0 mm internal diameter transmits 540.0 kW at 720.0 rpm, and is subjected to an axial thrust of 60.0 kN. Find the safe B.M. on the shaft, if the stresses should not exceed 70.0 MPa.

Solution VII The shaft transmits 540.0 kW at 720.0 rpm; the torque developed in the shaft is thus

$$T = (60P)/(2\pi n)$$

$$= (60 \times 540.0 \times 10^3)/(2 \times \pi \times 720) = 7\ 162.0 \text{ Nm}$$

The geometric properties of the shaft are given by

$$A = (\pi/4)(D_o^2 - D_i^2) = 5.027 \times 10^3 \text{ mm}^2$$

$$J = (\pi/32)(D_o^4 - D_i^4) = 8.545 \times 10^6 \text{ mm}^4$$

$$Z = [\pi/(32D_o)](D_o^4 - D_i^4) = 85.451 \times 10^3 \text{ mm}^3$$

The maximum shear stress in the shaft is given by

$$\tau_{max} = T(D_o/2)/J = 7\ 162.0 \times 10^3 \times 50.0/(8.545 \times 10^6) = 41.91 \text{ MPa}$$

The axial stress in the shaft is given by

$$\sigma_a = 60.0 \times 10^3/(5.027 \times 10^3) = 11.94 \text{ MPa}$$

From the given condition, the major principal stress should not exceed 70.0 MPa in the shaft; accordingly, we have

$$\sigma_1 \le 70.0 \le (\sigma_{max}/2) + [(\sigma_{max}/2)^2 + \tau_{max}^2]^{0.5}$$

$$\le (\sigma_{max}/2) + [(\sigma_{max}/2)^2 + 41.91^2]^{0.5}$$

Solving the above equation, we obtain

$\sigma_{max} \leq 44.91$ MPa

Net allowable bending stress on the section

$$\sigma_b = 44.91 - 11.94 = 32.97 \text{ MPa}$$

Maximum allowable B.M.

$$M = \sigma_b Z = 32.97 \times 85.451 \times 10^3 = 2.82 \text{ kNm}$$

The safe B.M. on the shaft for the given parameters is thus 2.82 kNm.

8.4 Compound shafts

The shafts analysed so far comprised a single material. In some cases, for economy or corrosion resistance, shafts may be fabricated of more than one material. Such shafts may comprise a core and an external shell as indicated in Fig.8.7. Such shafts are also known to be joined in parallel (in analogy to electric circuits); the shafts may also be coupled along their lengths, and are said to be in series. In all these cases, the shafts are assumed to be rigidly joined (deformations are equal at the contact surfaces).

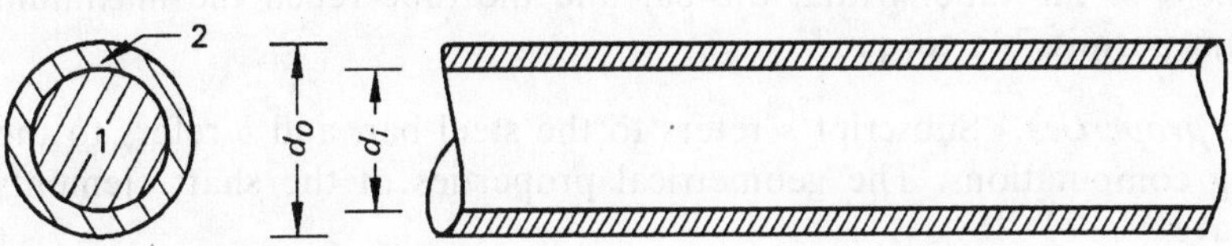

Fig.8.7 Composite shaft

The parameters of the core in Fig.8.7 are denoted by the subscript 1, while those of the tube are denoted by subscript 2. The shaft is statically indeterminate; the compatibility condition should be applied at the interface in order to analyse the system. When the composite shaft shown in Fig.8.7 is subjected to torsion, both the elements (inner core and the outer tube) are subjected to the same deformation all along their lengths. Applying Eqn.8.4, we have

$$\theta_1 = \theta_2 = \theta \quad \text{or} \quad T_1 l_1/(G_1 J_1) = T_2 l_2/(G_2 J_2)$$

Simplifying the above equation, we obtain

$$T_1 = T_2(l_2/l_1)[(G_1 J_1)/(G_2 J_2)] \qquad [8.12]$$

If the core and the tube are of the same length, we have

$$T_2 = T_1[(G_2 J_2)/(G_1 J_1)] \qquad [8.13]$$

From the equilibrium condition, we have

$$T = T_1 + T_2 \qquad [8.14]$$

Substituting for the value of T_2 from Eqn.8.13 in the above equation and simplifying, we obtain

$$T_1 = T(G_1 J_1)/(G_1 J_1 + G_2 J_2) \qquad [8.15]$$

In general, if the shaft comprises n elements, the torsional component sustained by element k can be expressed as

$$T_k = T(G_k J_k)/\sum(G_i J_i) \qquad [8.16]$$

The angle of twist can be obtained as

$$\theta = Tl/\sum(G_i J_i) \qquad [8.17]$$

The application of the above equations is illustrated in the following examples.

Problem VIII A brass tube of 1.0 mm thickness is shrunk over a mild steel bar of 25.0 mm diameter of the same length. Determine the increase in the torque resistance of the tube assuming no slip at the interface of the assembly. The eleastic properties are G_b= 33.4 GPa and G_s = 76.9 GPa, and the maximum allowable stresses are τ_b= 40.0 MPa and τ_s=70.0 MPa. What should be the thickness of the tube so that the bar and the tube reach the maximum stresses under the maximum permissible torque?

Solution VIII *Geometric properties* Subscript s refers to the steel bar, and b refers to the brass tube in the following computations. The geometrical properties of the shaft elements are determined first; we have

$$J_s = \pi D^4/32 = \pi \times 25.0^4/32 = 38\ 349.52\ \text{mm}^4$$

$$J_b = \pi D^3 t/4 = \pi \times 26.0^3 \times 1.0/4 = 13\ 804.16\ \text{mm}^4$$

Maximum allowable torque The maximum permissible torque on the brass tube is given by

$$T_b = \tau_b J_b/r_b \qquad \text{[Eqn.8.4]}$$

$$= 40.0 \times 13\ 804.16/(26.0/2) = 42\ 474.3\ \text{Nmm}$$

The maximum permissible torque on the steel bar is given by

$$T_s = \tau_s J_s/r_s \qquad \text{[Eqn.8.4]}$$

$$= 70.0 \times 38\ 349.52/(25.0/2) = 214\ 757.3\ \text{Nmm}$$

The maximum allowable moment on the compound shaft can be obtained from Eqn.8.15; we have

$$T = T_s[1.0 + (G_b J_b)/(G_s J_s)]$$

$$= 214\ 757.3 \times [1.0 + (33.4 \times 13\ 804.16)/(76.9 \times 38\ 349.52)] = 248\ 332.5\ \text{Nmm}$$

The torque component of the brass tube is given by

$$T_b = T - T_s = 248\ 332.5 - 214\ 757.3 = 33\ 575.2 \text{ Nmm}$$

This component is less than the maximum allowable torque on the tube computed above. Thus, the maximum allowable torque on the compound shaft is 248.33 Nm. The increase in the capacity of the brass bar is about 485 percent.

Tube thickness for the given criterion Applying the compatibility condition, we have

$$[T_s/(G_s J_s)] = [T_b/(G_b J_b)] = [(\tau_b J_b/r_b)]/(G_b J_b) = [\tau_b/(G_b r_b)]$$

or $\quad r_b = [\tau_b G_s J_s/(G_b T_s)] = [40.0 \times 76.9 \times 38\ 349.52/(33.4 \times 214\ 757.3)] \text{ mm} = 16.45 \text{ mm}$

The required tube thickness is $(16.45 - 12.50) = 3.95$ mm.

Problem IX A compound shaft comprises a steel bar of 60.0 mm diameter and 1.0 m length coupled rigidly to a tube of 40.0 mm internal diameter and 0.50 m length. Determine the maximum torque the shaft can sustain, if the allowable (i) stress is 120.0 MPa, and (ii) twist is 1 degree in 10 diameters. Adopt G = 76.9 GPa.

Solution IX *Geometric properties* Subscript b refers to the steel bar and t refers to the tube.

$$J_b = \pi D^4/32 = \pi \times (60.0)^4/32 = 1.272 \times 10^6 \text{ mm}^4$$

$$J_t = \pi(D_o^4 - D_i^4)/32 = \pi[(60.0)^4 - (40.0)^4]/32 = 1.021 \times 10^6 \text{ mm}^4$$

(i) Maximum allowable stress The tube will reach the allowable stress first because of the lower polar moment of interia (J); we have

$$T_t = \tau J_t/r_t = 120.0 \times 1.021 \times 10^6/30.0 = 4.084 \text{ kNm}$$

The allowable torque in the shaft is 4.084 kNm.

(ii) Maximum allowable twist The twists of the individual components have to be computed, and added together to check for the second criterion. The allowable twist

$$\begin{aligned} \theta &= [1/(10 \times 60.0)] \times 1\ 500.0 = 2.5° = 0.043\ 6 \text{ radian} \\ &= (T/G)[(l_b/J_b) + (l_t/J_t)] \qquad \text{[Eqn.8.4]} \\ &= [T/(76.9 \times 10^3)] \times [\{1\ 000.0/(1.272 \times 10^6)\} + \{500.0/(1.021 \times 10^6)\}] \end{aligned}$$

Solving the above equation, we obtain the allowable torque T=2.628 kNm.

8.5 Summary

Several structural elements such as machine shafts, beams and pipelines develop torsional moments under loading. The assumptions and theory of torsion are explained along with the analysis for stresses due to torsion in circular shafts. The effects of torsion combined with bending are also investigated. Some of the significant relationships are summarised here.

Torsion equation $[T/J] = [\tau/r] = [G\theta/l]$

Polar M.I.

Solid circular section $J = \pi D^4/32$

Hollow circular section $J = \pi (D_o^4 - D_i^4)/32$

Thin circular tube $J = \pi D^3 t/4$

Torsional modulus Circular sections $C = J/R$

Strain energy due to torsion $U = [T^2 l/(2GJ)]$

Power transmitted by shafts $P = 2\pi n T/60$

Equivalent B.M. $M_e = [M + \{M^2 + T^2\}^{0.5}]/2$

Equivalent torque $T_e = [M^2 + T^2]^{0.5}$

8.6 Exercises

8.1 Discuss the assumptions and limitations of torsion equation.

8.2 What is the difference in the behaviour of circular and non-circular shafts subjected to torsion?

8.3 Develop an expression for strain energy in a shaft subjected to torsion, and show that the maximum strain energy in the shaft is twice the total strain energy.

8.4 Show that the strain energy of a shaft is proportional to the fourth power of its diameter.

8.5 Show that the strain energy of a solid circular shaft is twice that of a thin circular shaft of the same cross sectional area and length for the same material when subjected to the same torsional stress.

8.6 Determine the maximum stress and deformation of a shaft of 50.0 mm diameter and 3.8 m length subjected to a torque of 27.0 kNm. Determine the minimum thickness required for a hollow circular shaft of 50.0 mm internal diameter for the
(i) same maximum stress and
(ii) same maximum deformation as the solid shaft.
Assume G = 85.0 GPa for the material.

8.7 A shaft of length l, fixed at both the ends is subjected to an anti-clockwise torsion T at a distance a from one end. Determine the torque distribution in the shaft.

8.8 A 9.0 m long shaft is fixed at both its ends, and is subjected to a torque of 100.0 kNm at 2.7 m from one end. Determine the minimum size of the shaft, if the allowable shear stress is 75.0 MPa, and the maximum allowable twist is 1 degree in 15 diameters; assume $G = 65.0$ GPa.

8.9 A shaft of length l fixed at both the ends is subjected to an anti-clockwise torque T_1 at a distance a from one end, and a clockwise torque T_2 at a distance b from the other end. Develop the torque distribution diagram.

8.10 A 12.0 m long shaft fixed at both the ends is subjected to an anti-clockwise torque of 38.0 kNm at a distance 2.5 m from one end, and a clockwise torque of 20.0 kNm at a distance 3.5 m from the other end. Determine the minimum size of the shaft, if the allowable shear stress is 55.0 MPa, and the allowable twist is one degree in 20 diameters; assume $G = 75.0$ GPa.

8.11 A 7.2 m long prismatic shaft of 65.0 mm diameter, restrained at its ends, is subjected to a uniformly distributed torque of 22.5 kNm^{-1}. Determine the maximum stress and deformation of the shaft, if the rigidity modulus is 50.0 GPa.

8.12 Compute the maximum torque a hollow mild steel shaft of external diameter 80.0 mm and 5.0 mm thickness can sustain with a factor of safety of 1.8, if the elastic limit in tension is 350.0 MPa. Use the criterion of
(i) maximum shear stress, and
(ii) maximum strain energy.

8.13 Design a hollow circular shaft to sustain a torque of 65.0 kNm, if the external to internal diameter ratio is 1.2. The allowable shear stress is 95.0 MPa. Use the criterion of
(i) maximum shear stress, and
(ii) maximum strain energy.

8.14 Determine the maximum power transmitted at 270 rpm by a mild steel shaft of 35.0 mm diameter, if the allowable shear stress is 75.0 MPa and the angle of twist is not to exceed 1 degree in a length of 12 diameters. Assume $G = 50.0$ GPa.

8.15 Design a shaft to transmit 0.9 MW of power at 120 rpm. The stress in the shaft should not exceed 50.0 GPa, and the angle of twist should not be more than 1 degree in a length of 10 diameters. Assume $G = 65.0$ GPa for the material.

8.16 A flywheel of 9.0 kN weight and 200.0 mm radius of gyration is mounted on a 60.0 mm diameter shaft of 1.5 m length; the shaft is supported on a bearing at the other end. If the bearing freezes suddenly while transmitting 30.0 kW at 250.0 rpm, determine the maximum stress developed in the shaft.
[Hint: Equate the kinetic energy of the wheel to the strain energy of the shaft; kinetic energy of a wheel $= mv^2/2$, where m is the mass, v is the linear velocity [$(= 2\pi\, n\, r/60)$, r = radius gyration].

8.17 A shaft supported on bearings 1.5 m apart transmits 65.0 kW at 360 rpm through a flywheel of mass 100.0 kg. Design the shaft for a maximum stress of 90.0 MPa, if the maximum torque can be 25 percent more than the mean torque.

8.18 A propeller shaft of 120.0 mm diameter transmits 290.0 kW at 630.0 rpm, and is subjected to an axial thrust of 90.0 kN. Find the safe bending moment on the shaft, if the stresses should not exceed 70.0 MPa.

8.19 Design a propeller shaft to transmit 90.0 kW at 470.0 rpm, and subjected to an axial thrust of 120.0 kN and a bending moment of 10.0 kNm, if the maximum stress should not exceed 60.0 MPa.

8.20 A brass tube of 1.2 mm thickness is shrunk over a mild steel bar of 20.0 mm diameter of the same length. Determine the increase in the torque resistance of the tube assuming no slip at the interface of the assembly. The elastic properties are G_b = 40.0 GPa and G_s = 80.0 GPa, and the maximum allowable stresses are τ_b = 50.0 MPa and τ_s = 90.0 MPa. What should be the thickness of the tube so that the bar and the tube reach the maximum stresses under the maximum permissible torque?

8.21 A compound shaft comprises a steel bar of 50.0 mm diameter and 0.50 length coupled rigidly to brass bar of the same diameter but 1.20 m length. Determine the maximum torque the shaft can sustain, if the allowable
(i) stress is 120.0 MPa in steel and 65.0 MPa in brass and (ii) twist is 1 degree in 12 diameters. Adopt G = 80.0 GPa for steel and 40.0 for brass.

Answers to selected problems

Chapter 1

1.25 $E = 211.6$ GPa and $\sigma_w = 240.2$ MPa
1.26 f.o.s. = 1.72 and $P_a = 1.39$ MPa
1.27 $a = 182.6$ mm and $P_a = 415.8$ kN
1.28 $d = 40.0$ mm
1.29 $\sigma = 147.6$ mm and f.o.s. = 1.68
1.30 $P_a = 6.79$ kN
1.31 The stresses are 916.7, 707.4 and 480.0 MPa; $P_{\max} = 13.35$ kN
1.32 $d = 33.3$ mm (35.0 mm), allowable length = 1 173.6 mm
1.33 28 strands
1.34 Fig. E 1.1 (a) − 2.07 mm (b) − 0.80 mm (c) + 9.06 mm (d) + 0.30 mm
1.35 Fig. E 1.1 (b) 166.7 mm (c) 2 287.0 mm (d) 1 004.3 mm
1.36 Fig. E 1.1 (b) 11.55 mm and $x = 222.2$ mm
1.37 Fig. E 1.1 (d) 22.02 mm, and $x = 362.0$ mm
1.38 Fig. E 1.1 (e) 0.51 mm (f) 0.54 mm
1.39 $\delta = 4Pl/(\pi EDd)$
1.40 $\delta = wl^2/(2EA)$
1.41 $A(x) = Ae^{wxP/A}$
1.42 Fig. E 1.2 Forces in kN
(a) −77.1, +59.0 (b) −127.5, +97.5 (c) −22.7, +0.0 (d) −435.3, +332.9
1.43 Fig. E 1.3 (a) +1.18 kN in both wires (b) +0.66 kN in steel wire and +1.31 kN aluminium wire
1.44 Fig. E 1.4 Forces in kN
(a) 6.92 and 55.39 (b) 17.93 and 104.58 (c) 95.75 and 568.13 (d) 25.93 and 207.41
1.45 Fig. E 1.5 Forces in kN
(a) 100.0, 150.0 and 100.0 (b) 153.9, 192.3 and 153.9
(c) 71.4, 107.4 and 71.4 (d) 152.0, 116.0 and 32.0
1.46 Fig. E 1.5 Forces in kN
(a) 58.3, 233.3 and 58.3 (b) 93.8, 312.5 and 93.8 (c) 41.7, 166.7 and 41.7 (d) 120.0, 180.0 and 0.0
1.47 81.4 kN, 10.9 percent and 1.1 percent
1.48 (i) $\sigma_{al} = +13.4$ MPa and $\sigma_c = -5.0$ MPa (ii) $\sigma_{al} = -4.9$ MPa and $\sigma_c = +1.9$ MPa
1.49 Fig. E 1.6 Stresses in MPa
(a) (i) $\sigma_c = -58.9$ and $\sigma_s = -530.0$ (ii) $\sigma_c = +25.5$ and $\sigma_s = +229.70$
(b) (i) $\sigma_c = -426.6$ and $\sigma_s = -47.4$ (ii) $\sigma_c = +184.9$ and $\sigma_s = +20.5$
1.50 Fig. E 1.7 (a) − 17.6, − 39.7 and − 17.6 MPa (b) − 176.7, − 46.6 and −176.7 MPa
(c) − 9.2, − 35.0 and − 30.8 MPa (d) − 174.0, − 88.9 and − 137.1 MPa
1.51 Fig. E 1.7 (b) 1.18 mm
1.52 Fig. E 1.7 (a) − 29.3, − 16.5 and − 29.3 MPa (b) − 191.0, − 18.1 and − 191.0 MPa
(c) − 20.5, − 13.6 and − 40.9 MPa (d) − 205.3, − 34.7 and − 160.0 MPa
1.53 Fig. E 1.1 (e) − 75.0 mm^3 (f) + 72.9 mm^3
1.54 22.23 m^3
1.56 0.012 6 mm^3/m^2

Chapter 2

2.8

	Mild steel	Brass	Aluminium	Cast iron
Modulus of resilience (Nmm/mm^3)	0.144	0.036	0.026	0.008
Modulus of toughness (Nmm/mm^3)	83.9	25.5	12.0	1.2

2.9 44.39 Nmm and 147.81 Nmm
2.10 0.156 Nmm/mm^3, 71.09 Nmm/mm^3 and 4.02 kNm
2.11 2.9 m long bar; 3.6 m long bar
2.12 The impact resistance of the aluminium bar is 2.04 times that of the steel bar
2.13 7.56 MPa and 1.70 mm
2.14 (i) 75.3 MPa, 1.13 mm and 90.5 (ii) 191.8 MPa, 1.80 mm and 58.6
2.15 382.33 mm
2.16 374.81 mm
2.17 Bar diameter required (i) 95.37 mm (ii) 95.84 mm
2.18 (i) 35.21 kN (ii) 16.33 kN
2.19 The bar yields but does not break
2.20 401.1 MPa and 83.56 mm
2.22 212.3 MPa and 106.1 MPa, and 1.91 mm

Chapter 3

3.6 130.7 kN
3.7 10.0 mm and 1 333.33 MPa
3.8 20.2 mm
3.9 124.3 MPa
3.10 Factor of safety = 1.41
3.11 118.4 MPa
3.12 5.56 MPa and 2.025 MN

Chapter 4

4.10

	Principal stresses (MPa)		Principal planes (degrees)	
(i)	+ 512.4	+ 97.6	+ 37.3	+ 127.3
(ii)	+ 309.7	– 419.7	– 16.6	+ 73.4
(iii)	– 97.6	– 512.4	+ 52.7	– 37.3
(iv)	+ 419.7	– 309.7	+ 16.6	+ 106.6

	Stress (MPa) Inclination of plane (degrees) 45		60	
(i)	+ 105.5	+ 55.5	+ 159.3	+ 147.6
(ii)	– 255.0	– 305.0	– 380.7	– 164.1
(iii)	– 505.07	– 55.0	– 505.7	+ 52.4
(iv)	– 145.0	+ 305.0	+ 34.3	+ 364.1

4.11 $\sigma_1 = +164.9$ MPa, $\sigma_2 = -184.9$ MPa, $\theta_1 = -29.5°$ and $\theta_2 = +60.4°$
$\sigma_{-60} = +74.9$ MPa, $\tau_{-60} = +152.9$ MPa, $\sigma_{30} = -94.9$ MPa and $\tau_{30} = -152.9$ MPa
4.12 (i) 134.2 MPa (ii) 120.4 MPa
4.13 $\sigma_1 = +330.1$ MPa, $\sigma_2 = -530.1$ MPa and $\tau_{max} = 315.1$ MPa
4.14 $\tau_{max} = 210.2$ MPa

Chapter 5

5.5

Fig. No.	M_{max} (kNm)	M_{min} (kNm)	SF_{max} (kN)	SF_{min} (kN)
E 5.1 (a)	+ 3 100.1	0.0	+ 350.0	– 350.0
(b)	+ 847.2	0.0	+ 126.0	– 574.0
(c)	+ 580.0	0.0	+ 165.1	– 234.2
(d)	+ 450.0	0.0	+ 90.0	– 90.0
(e)	0.0	– 445.5	0.0	– 54.0
(f)	0.0	– 123.3	0.0	– 35.0
(g)	0.0	– 526.5	+ 83.1	0.0
(h)	+ 61.04	– 25.0	+ 30.5	– 25.7
(i)	+ 17.2	– 3.8	+ 11.1	– 12.3
(j)	+ 61.9	– 22.1	+ 37.4	– 37.4
(k)	+ 37.5	– 22.8	+ 11.8	– 33.2

Chapter 6

6.5 $a^4/24$, $\sqrt{2}a^3/24$
6.6 7.6 percent
6.7 (i) 10.7 percent (ii) 99.9 percent
6.8 One ISWB 450 with 300.0 × 10.9 mm thick plates requires 17.7 percent less material than two ISWB 450 for the same moment capacity, hence is more economical
6.9 ± 86.74 MPa
6.10 $\sigma_t = +\ 2.13$ MPa and $\sigma_b = -\ 3.83$ MPa; 42.24 kN/m
6.11 157.38 kNm
6.12 (i) 3.93 m (ii) 7.86 m

Chapter 7

7.5 $\sigma_{min} = -\ 22.1$ MPa, $\sigma_{max} = +\ 10.6$ MPa and $P = 164.4$ kN
7.6 113.6 mm
7.7 Circle of radius $(D/4)$
7.8 Parallelogram bounded by one third the side length from the centroid
7.9 $\sigma_{max} = +\ 229.2$ MPa, $\sigma_{min} = 25.5$ MPa, $E = 98.5$ GPa, – 13.8 percent
7.10 Parallelogram with semidiagonals of 106.2 and 70.1 mm
7.11 $\sigma_{max} = +\ 5.75$ MPa and $\sigma_{min} = -\ 13.75$ MPa

Chapter 8

8.6 $\tau_{max} = 110.0$ MPa and $\theta = 11.3°$ (i) 5.52 mm (ii) 4.73 mm
8.7 Torsional moment at the origin $= T(l-a)/l$ and Ta/l at the other end
8.8 211.25 mm
8.9 Torsion in the segments from the left are $[T_1(l-a) - T_2 b]/l$, $-[T_1 a + T_2 b]/l$ and $T_2 - [T_1 a + T_2 b]/l$
8.10 130.95 mm and 169.96 mm
8.11 166.91 mm and 10.59°
8.12 (i) 4.04 kNm (ii) 5.22 kNm
8.13 (i) 188.91 and 157.33 mm (ii) 175.72 and 146.44 mm
8.14 8.66 kW **8.15** 90.0 mm
8.16 644.3 MPa **8.17** 223.3 mm
8.18 11.47 kNm **8.19** 126.6 mm
8.20 348.7 percent, $t = 1.21$ mm
8.21 (i) 1.595 kNm (ii) 0.837 kNm

Glossary

Axis of member is the line joining the centroids of the cross sections of a member along the length. A member is idealised as a line, denoted by the axis, in elementary analysis.

Beam is a member subjected to loads transverse (normal) to its axis; it is subjected predominantly to bending moments.

Bending moment at a section is the moment of all the forces on one side of the section about the section. It is generally referred to as moment for brevity, but is different from moment as explained later, and is usually expressed in kNm.

Bulk modulus (volumetric modulus) is the ratio of volumetric stress (applied equally in all directions) to volumetric strain within the elastic limit of the material. It is generally expressed in GPa.

Column is a structural member subjected predominantly to axial forces.

Compatibility is the relationship between the displacements at a section (usually a joint) in a body (beam, column or frame).

Deformation is the change in the dimensions of a body. It is positive, if the dimension increases (tension), and negative, if the body is compressed. Deformations may occur due to applied loads or temperature changes.

Displacement of a body is the movement from its previous (original) position. It may be due to applied loads or temperature changes.

Ductility is the ability of the material to be drawn into thin wires without breaking. It is generally expressed as the percentage strain at fracture.

Elasticity is the ability of a material to regain its original shape (and dimensions) completely on removal of the applied load.

Equilibrium of forces implies that their resultant is zero, and the body on which they act remains static.

Factor of safety is the numerical value with which the yield (proof) strength of a material is divided in order to obtain its working strength.

Force is the entity that changes the state of a body; the change may be deformation, displacement or motion.

Free body diagram is an isolated part of a body (beam or frame), with the external forces and the forces acting upon it due to the removed part (or parts) noted.

Lever arm is the perpendicular distance between the direction of the force and the point about which the moment is computed.

Moment of a force about a point (or section) is the product of the force and its lever arm. (Moment of the forces acting on a body in equilibrium about any point within or outside the body is zero, while bending moment at a section may or may not be zero).

Moment of inertia is the integral $I = \int y^2 dA$ over the cross sectional area about the specified axis. It is expressed in mm^4 or m^4.

Modulus of section is the ratio of the moment of inertia to the distance of the extreme fibre from the neutral axis ($Z = I/y_{max}$). It is expressed in mm^3 or m^3.

Neutral axis is the line in the beam cross-section along which strains (and, hence, stresses) are zero.

Poisson's ratio is the ratio of the lateral strain to the longitudinal strain ($\nu = -\varepsilon_t/\varepsilon$).

Rigidity modulus (shear modulus) is the ratio of shear stress to the corresponding shear strain within the elastic limit of the material. It is generally expressed in GPa.

Shear force at a section is the unbalanced force, normal to the beam axis, on one side of the section (to the right or left).

Strain is the ratio of deformation of the body to its corresponding dimension (or deformation per unit length of the body). It is termed axial strain (under axial force), shear strain (under shear force), bending strain (due to flexure) or volumetric strain (under hydrostatic pressure).

Strain energy is the work done by the internal forces induced in a body (internal energy). It is equal to the work done by the applied (external) forces on the body, and is released when the load is removed.

Stress is the force per unit area on which the force acts. It is usually expressed in MPa, and termed axial stress (under axial force), shear stress (under shear force), bending stress (due to flexure) or volumetric stress (under hydrostatic pressure).

Support is the point of contact of a body with a larger body. A body transfers external forces through axial or flexural deformations to another body through points of contact (supports).

Torsion is the force that twists a member (or a moment applied normal to member axis). It is usually expressed in kNm.

Working strength (or stress) of a material is the maximum allowable (or permissible) value of stress during the service life of a system (machine or structure).

Yield strength (or yield stress) of a material is the stress beyond which the material develops excessive strains (deformations).

Young's modulus (or elastic modulus) is the ratio of axial stress to the corresponding strain within the elastic limit of the material. It denotes the stress per unit strain of the material, and is expressed in GPa.

References

1. P. Bhatt and H.M. Nelson, *Marshall and Nelsons structures*, pp. 602, The English Language Book Society, London, 1990.

2. S.H. Crandall, N. C. Dahl and T. J. Lardner, *An introduction to mechanics of solids*, Second Edition, pp. 624, McGraw-Hill International Book Company, New York, 1978.

3. W. A. Nash, *Theory and problems of strength of materials*, Fourth Edition, pp. 594, Schaums Outline Series, McGraw-Hill International Book Company, New York, 1987.

4. D. S. Prakash Rao, *Strength of materials — a practical approach*, Volume I, pp. 646, Universities Press, Hyderabad, 1999.

5. A. Pytel and F. L. Singer, *Strength of materials*, Fourth Edition, pp. 594, Harper & Row, New York, 1987.

6. G. H. Ryder, *Strength of materials*, pp. 340, The English Language Book Society, London, 1986.

7. S. P. Timoshenko, *Strength of materials*, Volume 1, pp. 376, CBS Publishers & Distributors, New Delhi, 1986.

Index

Allowable stress, 18
Arbitrary section, 120
Axial force, 6, 10
Axial thrust, 146

Beams of uniform strength, 127
Bending equation, 115
Bending moment, 6, 95
Bending moment diagram (B.M. diagram), 97, 98
Bending theory, 112
Bi-axial stress, 79
Built-up section, 124
Bulk modulus, 41–44
Butt joints, 68

Cantilever, 94, 106–109
Circular ring, 119
Circular section, 118, 126
Circular section (middle quarter rule), 134
Coefficient of thermal expansion, 34, 44
Combined stresses, 129
Combined torsion and bending, 146
Compatibility conditions, 27, 35
Complementary shear, 69, 70, 144
Complementary shear stress, 137
Composite shaft, 149, 151
Compound bar(s), 27, 31, 35, 36
Compound shafts, 149
Convention(s) and Notation, 2, 74, 95
Coordinate system, 3
Core, 133
Cotter pin, 69

Deformations, 4
Direct shear, 67, 68
Double shear, 68, 72
Ductility, 16, 17

Eccentricity, 129, 130
Elastic limit, 13
Elastic modulus, 125
Elastic range, 17
Energy, 51
Energy principle, 54
Engineer's bending theory, 112
Engineer's strain, 14
Engineer's stress, 14

Equilibrium, 6
Equilibrium condition, 79, 80
Equivalent B.M., 147, 152
Equivalent force, 61
Equivalent load, 56, 64
Equivalent torque, 152
Equivalent torsion, 147
Euler–Bernoulli's bending equation, 113, 115, 128
Euler—Bernoulli's theory, 113
External work, 51–53

Factor of safety, 18, 19
Failure stress, 14
Fixed support, 4
Flexural analysis, 112
Flexural shear, 67
Flexural stress, 112
Free body diagram(s), 9, 27, 75, 94, 100, 102

Gage length, 13, 15
General stress analysis, 81

High strength steel, 14
Hinged support, 4
Hogging bending moment, 96
Hollow circular section, 119, 140
Hollow rectangular section, 118, 126
Hooke's law, 12, 13, 42, 44
Hookean materials, 112

Impact factor, 58
Impact load analysis, 55
Impact loading, 51, 58
Indeterminate systems, 27
Internal work, 52, 53, 55

Kern, 132, 133, 135
Kinetic energy, 57, 63

Lap joint, 67
Leaf springs, 127
Lower yield point, 13

Major principal stress, 83
Middle third rule, 133
Mild steel, 13, 14
Minor principal stress, 83
Modulus of elasticity, 12

Mohr's circle, 87, 88, 90, 91
Mohr's stress circle, 74, 86, 87
Moment of inertia, 114, 115, 120, 128

Necking, 14
Negative bending moment, 96
Negative S.F., 95
Neutral axis, 113, 114, 118
Nominal strain, 14
Nominal stress, 14
Non-prismatic bar, 25
Non-prismatic members, 24
Non-prismatic shafts, 145
Normal forces, 6

Overhang, 110

Plastic, 13
Plastic range, 17
Poisson's effect, 40
Poisson's ratio, 39, 40, 43, 44
Polar moment of inertia (M.I.), 139, 152
Positive bending moment, 96
Positive S.F., 95
Principal plane(s), 82, 87, 90
Principal stress, 82–84, 87, 90, 91
Principle of superposition, 18, 129
Product of inertia, 115
Proof stress, 14, 15
Proportionality limit, 13
Punching shear, 69
Pure shear, 67
Pure shear stress, 78

Rectangular section, 117, 126, 130, 132
Resilience modulus, 53, 54, 58, 64
Rigidity modulus, 71, 72, 85
Riveted joints, 67
Roller support, 4

Sagging bending moment, 96
Second moment of area, 114
Section modulus, 124–128
Shaft, 137, 141, 142
Shear force, 67, 95
Shear force diagram (S.F. diagram), 97, 98
Shear modulus, 71
Shear strain, 70, 72
Shear stress, 67, 69, 72, 74, 87
Simple beam, 93, 99, 101–106
Simple bending theory, 112
Simply supported beam, 93
Single shear, 68, 72
Solid circular section, 139
Solid circular shafts, 147
Stepped bar, 22
Strain, 11, 13, 43, 44
Strain energy, 52, 53, 55, 56, 59, 60, 63, 64, 72, 145, 146, 152
Strain hardening, 13
Stress, 10, 11, 43, 44
Stress analysis, 74
Stress–strain, 13
Stress–strain curve, 15
Supports, 4
System of units, 1

Theorem of parallel axes, 115, 116, 128
Theorem of perpendicular axes, 116, 128
Thermal strain, 12, 34
Thermal stress(s), 34–36
Thin circular tube, 140
Torque, 6, 142, 145, 146
Torsion, 137
Torsion analysis, 137
Torsion equation, 138, 152
Torsional modulus, 152
Torsional moment, 6
Torsional rigidity, 139
Torsional shear, 67
Torsional stiffness, 139
Torsional stresses, 137
Toughness modulus, 54, 58, 64
Transverse forces, 6
Triangular section, 118, 126

Ultimate stress, 13
Uniformly distributed load (u.d.l.), 3
Uniformly distributed torque, 143, 144
Uniformly varying load, 3
Uniform stress, 26
Upper yield point, 13

Varying torque, 143
Volumetric strain, 41–43
Volumetric stress, 42

Warping, 138
Work, 51
Working stress, 18

Young's modulus, 12, 13, 43